MAMMAL

Fruit bat crawling

Bennett's wallaby

Hare skeleton

Red fox

Senegal bushbaby

Gerbil carrying nesting material

EYEWITNESS GUIDES

MAMMAL

Written by
STEVE PARKER

Hazelnuts
opened
by dormouse

Pine cone
chewed by
squirrel

Chimpanzee
clapping

Hedgehog about to unroll

Puppies playing

Hedgehog
footprint

Antelope horn

Lower jawbone

Chinchilla eating nuts

DK

DORLING KINDERSLEY • LONDON
in association with
THE NATURAL HISTORY MUSEUM • LONDON

Domestic cat

Badger rolling

Lion's tail bob

Baby wallaby

Rabbit sniffing the air

Black hooded rat

DK

A DORLING KINDERSLEY BOOK

This Eyewitness ® Guide has been
conceived by Dorling Kindersley Limited
and Editions Gallimard

Project editor Sophie Mitchell
Art editor Neville Graham
Managing editor Vicky Davenport
Managing art editor Jane Owen
Special photography
Jane Burton and Kim Taylor,
Dave King and Colin Keates
Editorial consultants
The staff of the Natural History Museum, London

First published in Great Britain in 1989
by Dorling Kindersley Limited,
9 Henrietta Street, London WC2E 8PS

Reprinted 1989, 1990, 1991, 1992, 1993, 1997

Visit us on the World Wide Web at
http://www.dk.com

British Library Cataloguing in Publication Data
Parker, Steve
  Mammal
  1. Mammals
  I. Title II. Series
  599

ISBN 0-86318-340-9

Colour reproduction by Colourscan, Singapore
Typeset by Windsorgraphics, Ringwood, Hampshire
Printed in Singapore by Toppan Printing Co. (S) Pte Ltd.

Contents

Golden hamster carrying baby

The mammal world

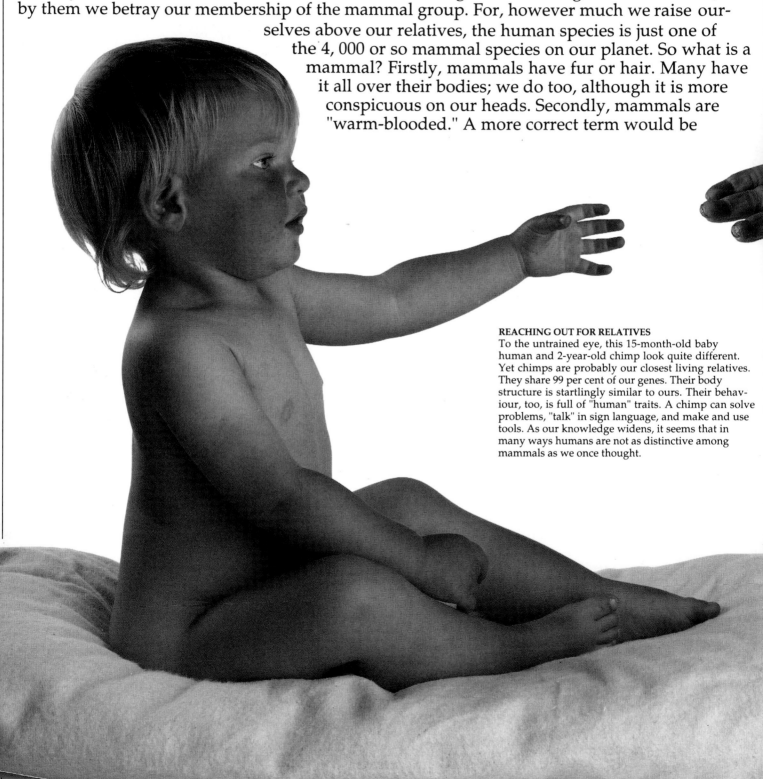

Humans are only one of perhaps 10 million different species, or types, of animals in the world. With some animals we may feel uneasy, even though there is no reason: a harmless snake, perhaps, or a slimy snail. Yet others seem to demand our interest. Bushbabies, seal-cubs, dolphins, kittens, and koalas - we are drawn by their furriness and their warm bodies, and the way the mother looks after her babies. These are things that we recognize in ourselves, and by them we betray our membership of the mammal group. For, however much we raise ourselves above our relatives, the human species is just one of the 4, 000 or so mammal species on our planet. So what is a mammal? Firstly, mammals have fur or hair. Many have it all over their bodies; we do too, although it is more conspicuous on our heads. Secondly, mammals are "warm-blooded." A more correct term would be

REACHING OUT FOR RELATIVES
To the untrained eye, this 15-month-old baby human and 2-year-old chimp look quite different. Yet chimps are probably our closest living relatives. They share 99 per cent of our genes. Their body structure is startlingly similar to ours. Their behaviour, too, is full of "human" traits. A chimp can solve problems, "talk" in sign language, and make and use tools. As our knowledge widens, it seems that in many ways humans are not as distinctive among mammals as we once thought.

"homoiothermic," meaning that a constant internal body temperature is maintained, usually above that of the environment, rather than adjusting to the temperature of the surroundings. In this way mammals can stay active even in cold conditions. Thirdly, mammals feed their young on milk. The milk is made in specialized skin structures called mammary glands, hence our group's biological name, Mammalia. This book sets out to explore the world of the mammals - what they look like, their body structure, evolution, breeding, habits, and behaviours - and in doing so will hopefully cast some light on our place in their world.

MAMMALS AS ANIMALS
There are about 4,000 species (types) of mammals. Because of the number of domestic species, and the popularity of mammals in zoos, we are more familiar with them than other animal groups. Yet there are some 9,000 species of birds, 20,000 species of fish, and 100,000 species of spiders and scorpions. All these pale against the greatest animal group - the insects, with at least 1 million species, and possibly 10 times that number.

Noah's Ark took on board two of every mammal species - one male and one female

Making sense of mammals

WE CAN APPRECIATE the beauty and wonder of mammals without knowing their scientific names or evolutionary origins. But a deeper understanding of body structure, behaviour, and evolution needs, like any aspect of science, a framework for study. This framework is provided by "taxonomy", the grouping and classifying of living things. Every living animal has a scientific name that is recognized across the world and in all languages. This avoids confusion, since local or common names vary from country to country, and even from place to place within the same country. Each kind of animal is known as a species. Species are grouped together into genera, genera are grouped into families, families into orders, and orders into classes . . . and this is where we can stop, since all mammals belong to one class, Mammalia. The following four pages show the skulls of representatives of the 20 or so main orders of living mammals, and list the types of animals that belong in each one. The coloured lines indicate their probable evolutionary relationships.

EDENTATES (Edentata)
Includes anteaters, tamanduas, armadillos, sloths.
About 30 species
Skull shown: Greater long-nosed armadillo
See also pp. 22, 27, 29, 51

 Armadillo

Monkey

MARSUPIALS OR POUCHED MAMMALS (Marsupialia)
Includes kangaroos, wallabies, wombats, possums, opossums, dunnarts, bandicoots, cuscuses.
About 270 species
Skull shown: Mountain cuscus
See also pp. 3, 4, 10, 20, 22, 27, 30-31

Kangaroo

MONKEYS AND APES (Primates)
Includes lemurs, bushbabies, lorises, pottos, tarsiers, marmosets, tamarins, monkeys, apes, humans.
About 180 species
Skull shown: Vervet monkey
See also pp. 2, 3, 6-7, 16-17, 21, 22-3, 29, 37, 38, 44, 49, 58

Pangolin

PANGOLINS (Pholidota)
Pangolins.
About 7 species
Skull shown: Chinese pangolin
See also p. 27

Platypus

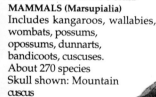

EGG-LAYING MAMMALS (Monotremata)
Platypuses, echidnas. Generally regarded as the most "primitive" mammals since they lay eggs (like reptiles) and do not give birth to formed young.
3 species
Skull shown: Platypus
See also pp. 16, 25, 27, 30, 56

INSECTIVORES (Insectivora)
Includes shrews, moles, golden moles, desmans, hedgehogs, moonrats, solenodons, tenrecs.
About 350 species
Skull shown: Greater moonrat
See also pp. 3, 24-5, 51, 57, 61

Shrew

Aardvark

AARDVARK (Tubulidentata)
1 species
Skull shown: Aardvark
See also p. 51

Kangaroo

Honey possum

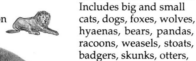

Lion

CARNIVORES (Carnivora)
Includes big and small
cats, dogs, foxes, wolves,
hyaenas, bears, pandas,
racoons, weasels, stoats,
badgers, skunks, otters,
mongooses, civets.
About 230 species
Skull shown: Egyptian
mongoose
See also pp. 2, 3, 4, 16-17,
21, 28-9, 34-9, 42-3, 46-7,
49, 50, 53, 56, 59, 60, 63

BOUND BY THE POUCH
The kangaroo and honey possum look
quite different, yet they are both marsupials.
The important common feature is the pouch in
which the baby suckles and grows after birth.
Only marsupials have this feature.

SEALS (Pinnipedia)
Seals, sea lions, walruses.
About 33 species
Skull shown: Grey seal
See also pp. 10, 20, 51, 59, 63

Seal

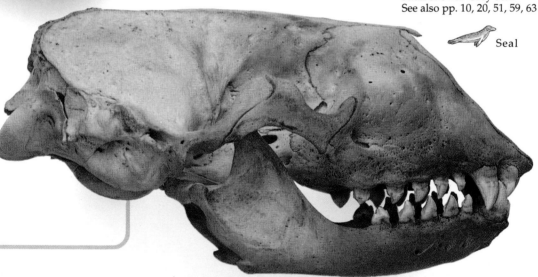

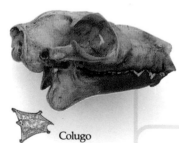

Colugo

BATS (Chiroptera)
Includes flying foxes (fruit bats), vampires,
and all other bats.
About 950 species (nearly one-quarter of
all mammal species)
Skull shown: Common flying fox
See also pp. 2, 18-19, 63

Bat

COLUGOS (Dermoptera)
Also called flying lemurs.
2 species
Skull shown: Malayan
colugo
See also p. 19

Rat

RODENTS (Rodentia)
Includes rats and mice, dormice,
gerbils, beavers, squirrels,
porcupines, chinchillas, pacas,
voles, hamsters, chipmunks.
About 1,700 species
Skull shown: Giant pouched rat
See also pp. 2, 4, 5, 16, 20, 22-3, 27,
32-3, 44-5, 48-9, 51, 52-3, 54-5, 61, 63

RABBITS AND HARES (Lagomorpha)
Includes rabbits, cottontails,
jackrabbits, hares, pikas.
About 45 species
Skull shown: European rabbit
See also pp. 2, 4, 60

Rabbit

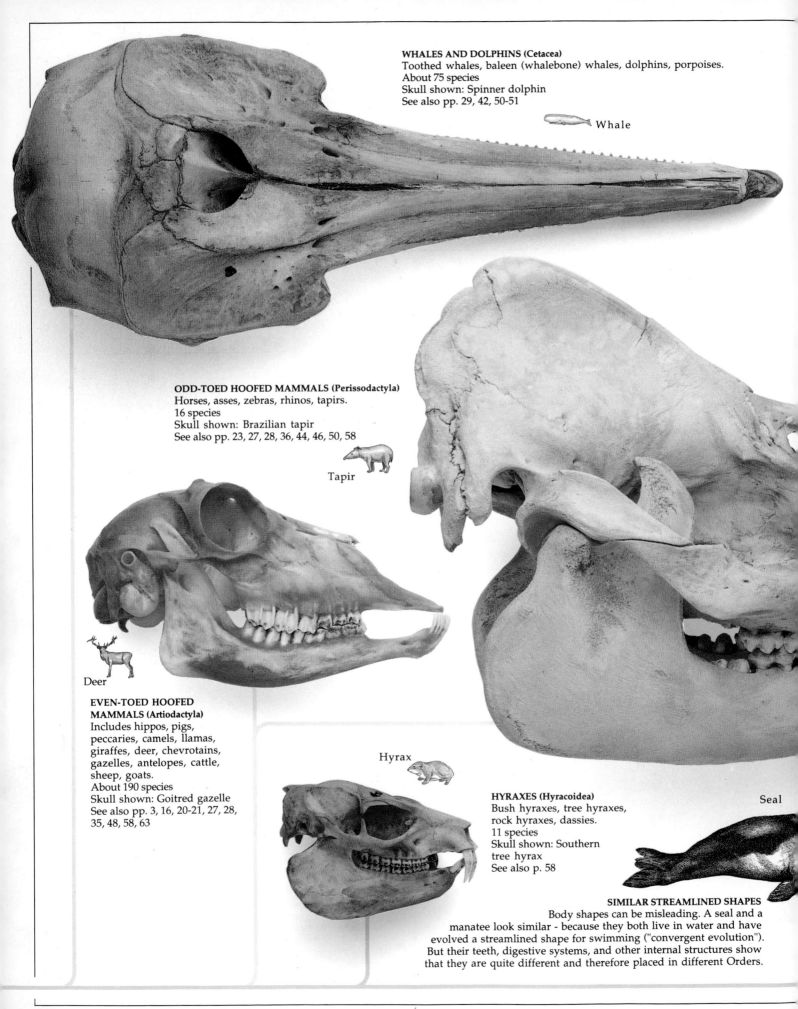

WHALES AND DOLPHINS (Cetacea)
Toothed whales, baleen (whalebone) whales, dolphins, porpoises.
About 75 species
Skull shown: Spinner dolphin
See also pp. 29, 42, 50-51

Whale

ODD-TOED HOOFED MAMMALS (Perissodactyla)
Horses, asses, zebras, rhinos, tapirs.
16 species
Skull shown: Brazilian tapir
See also pp. 23, 27, 28, 36, 44, 46, 50, 58

Tapir

Deer

EVEN-TOED HOOFED MAMMALS (Artiodactyla)
Includes hippos, pigs, peccaries, camels, llamas, giraffes, deer, chevrotains, gazelles, antelopes, cattle, sheep, goats.
About 190 species
Skull shown: Goitred gazelle
See also pp. 3, 16, 20-21, 27, 28, 35, 48, 58, 63

Hyrax

HYRAXES (Hyracoidea)
Bush hyraxes, tree hyraxes, rock hyraxes, dassies.
11 species
Skull shown: Southern tree hyrax
See also p. 58

Seal

SIMILAR STREAMLINED SHAPES
Body shapes can be misleading. A seal and a manatee look similar - because they both live in water and have evolved a streamlined shape for swimming ("convergent evolution"). But their teeth, digestive systems, and other internal structures show that they are quite different and therefore placed in different Orders.

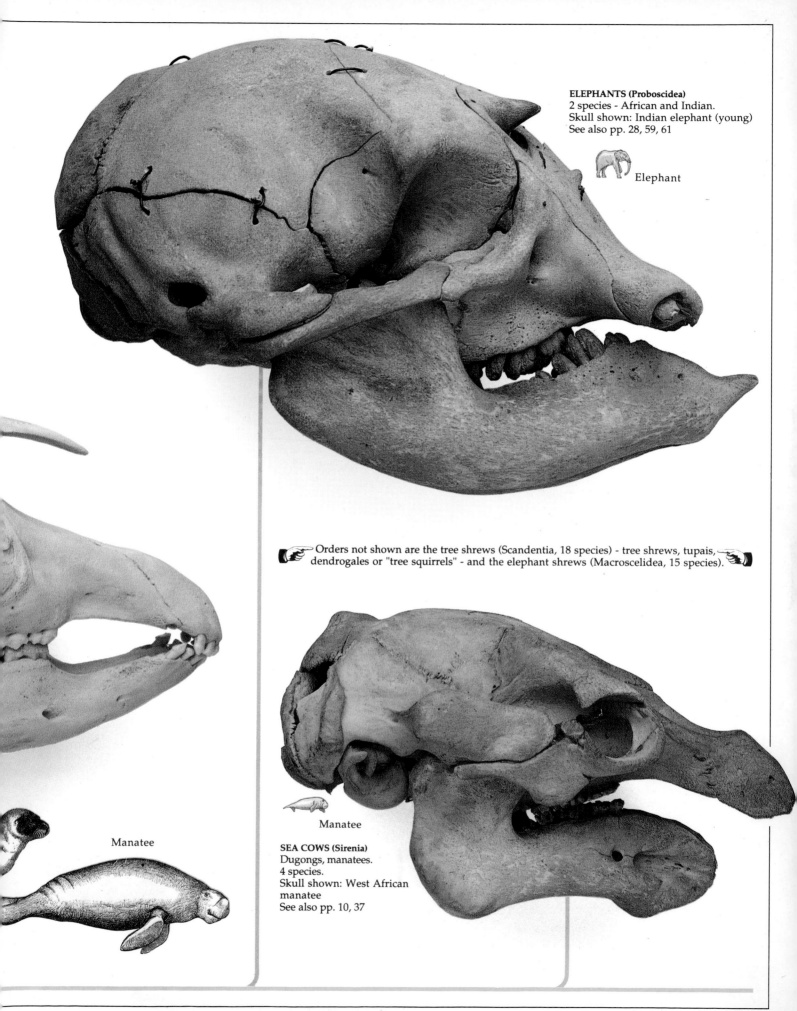

ELEPHANTS (Proboscidea)
2 species - African and Indian.
Skull shown: Indian elephant (young)
See also pp. 28, 59, 61

Elephant

☞ Orders not shown are the tree shrews (Scandentia, 18 species) - tree shrews, tupais, ☜
dendrogales or "tree squirrels" - and the elephant shrews (Macroscelidea, 15 species).

Manatee

Manatee

SEA COWS (Sirenia)
Dugongs, manatees.
4 species.
Skull shown: West African
manatee
See also pp. 10, 37

The evolution of mammals

An early rhino?

As FAR AS WE KNOW, mammals appeared on the Earth some 200 million years ago. We "know" because we have found their fossils: bones, teeth, and other parts that have been turned to stone and preserved in rocks. Since some of the features we recognize in living mammals (warm blood, fur, and milk) do not fossilize we must look for other clues. These must be based on the bones. So another two important features for being a mammal, alive or fossilized, are a particular kind of jaw (one bone in each lower jaw, not several like the reptiles), and tiny bones in the middle ear cavity. Mammals did not exactly burst upon the evolutionary scene. During their first 100 million or so years, life on land was mainly dominated by huge dinosaurs, while pterosaurs flew above and ichthyosaurs swam in the sea. The first true mammals were probably small, shrew-like creatures that lived nocturnally and fed by eating insects and stealing dinosaur eggs. As the dinosaurs died out, and finally disappeared some 65 million years ago, mammals filled their place.

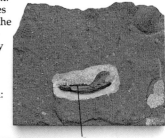

Skull from above

Lower jaw

MAMMAL ANCESTOR?
The cynodonts were "mammal-like reptiles" of the Triassic period. Their teeth were not all the same, as in other reptile groups, but were different shapes, specialized to do certain jobs. This is one of the mammal's characteristics, although some modern species (such as the dolphins) have re-evolved teeth that are all the same, in response to their diet (p. 51). Species shown: *Thrinaxon liorhinus* (S. Africa)

ONE OF THE FIRST
Embedded in rocks laid down in the middle Jurassic, in what is now England, is this jaw of a triconodont. These creatures were some of the earliest mammals, and they were rat- to cat-sized predators. Species shown: *Phascolotherium bucklandi* (Oxfordshire, UK)

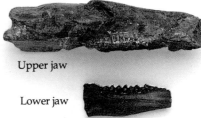

Lower jaw embedded in rock

Upper jaw

Lower jaw

SUCCESSFUL LINE
Freed from the domination of the dinosaurs, mammals changed rapidly during the Palaeocene and Eocene as evolution "experimented" with new forms. Some died out. This one did, but its general design was continued to the present day. It was an early relative of the horse, from the Eocene period. Species shown: *Hyracotherium vulpiceps* (Essex, UK)

THE STAGE IS SET
Into this sort of world, populated by giant fern-like plants, fish, insects, and reptiles, the first mammals emerged some 200 million years ago.

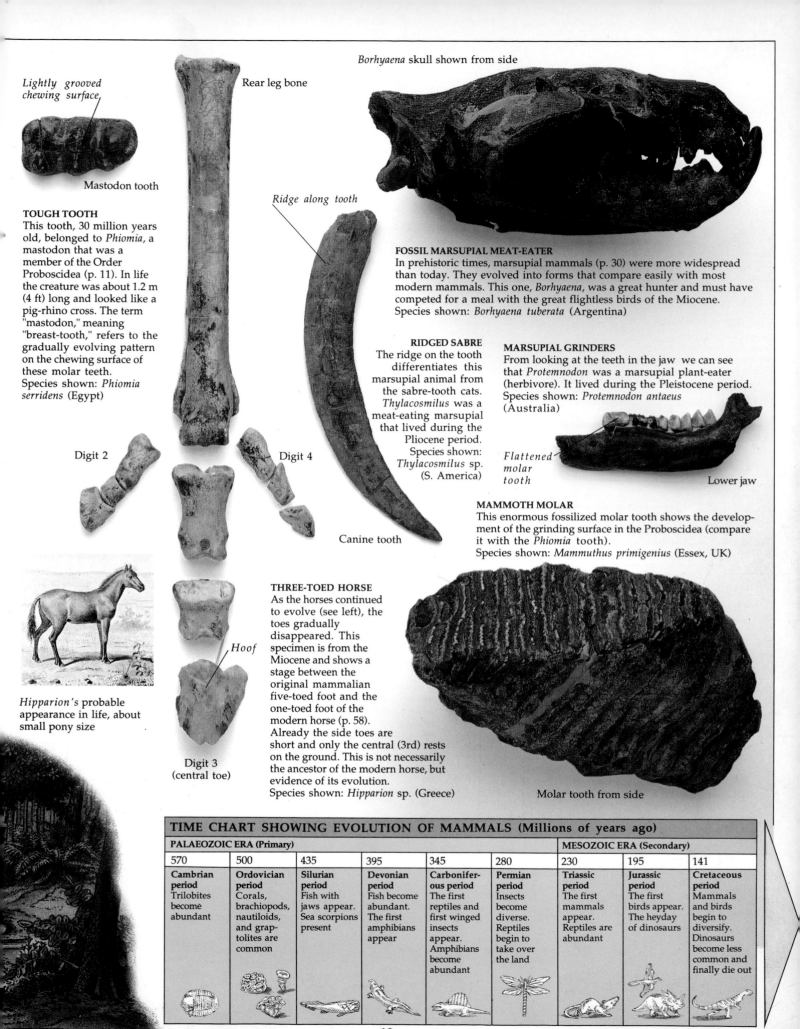

Lightly grooved chewing surface

Mastodon tooth

Rear leg bone

Borhyaena skull shown from side

Ridge along tooth

TOUGH TOOTH
This tooth, 30 million years old, belonged to *Phiomia*, a mastodon that was a member of the Order Proboscidea (p. 11). In life the creature was about 1.2 m (4 ft) long and looked like a pig-rhino cross. The term "mastodon," meaning "breast-tooth," refers to the gradually evolving pattern on the chewing surface of these molar teeth.
Species shown: *Phiomia serridens* (Egypt)

Digit 2

Digit 4

FOSSIL MARSUPIAL MEAT-EATER
In prehistoric times, marsupial mammals (p. 30) were more widespread than today. They evolved into forms that compare easily with most modern mammals. This one, *Borhyaena*, was a great hunter and must have competed for a meal with the great flightless birds of the Miocene.
Species shown: *Borhyaena tuberata* (Argentina)

RIDGED SABRE
The ridge on the tooth differentiates this marsupial animal from the sabre-tooth cats. *Thylacosmilus* was a meat-eating marsupial that lived during the Pliocene period.
Species shown: *Thylacosmilus* sp. (S. America)

MARSUPIAL GRINDERS
From looking at the teeth in the jaw we can see that *Protemnodon* was a marsupial plant-eater (herbivore). It lived during the Pleistocene period.
Species shown: *Protemnodon antaeus* (Australia)

Flattened molar tooth

Lower jaw

MAMMOTH MOLAR
This enormous fossilized molar tooth shows the development of the grinding surface in the Proboscidea (compare it with the *Phiomia* tooth).
Species shown: *Mammuthus primigenius* (Essex, UK)

Canine tooth

Hipparion's probable appearance in life, about small pony size

Hoof

THREE-TOED HORSE
As the horses continued to evolve (see left), the toes gradually disappeared. This specimen is from the Miocene and shows a stage between the original mammalian five-toed foot and the one-toed foot of the modern horse (p. 58). Already the side toes are short and only the central (3rd) rests on the ground. This is not necessarily the ancestor of the modern horse, but evidence of its evolution.
Species shown: *Hipparion* sp. (Greece)

Digit 3 (central toe)

Molar tooth from side

TIME CHART SHOWING EVOLUTION OF MAMMALS (Millions of years ago)

PALAEOZOIC ERA (Primary)						MESOZOIC ERA (Secondary)		
570	500	435	395	345	280	230	195	141
Cambrian period Trilobites become abundant	**Ordovician period** Corals, brachiopods, nautiloids, and graptolites are common	**Silurian period** Fish with jaws appear. Sea scorpions present	**Devonian period** Fish become abundant. The first amphibians appear	**Carbonifer-ous period** The first reptiles and first winged insects appear. Amphibians become abundant	**Permian period** Insects become diverse. Reptiles begin to take over the land	**Triassic period** The first mammals appear. Reptiles are abundant	**Jurassic period** The first birds appear. The heyday of dinosaurs	**Cretaceous period** Mammals and birds begin to diversify. Dinosaurs become less common and finally die out

The mammals diversify

Canine teeth (p. 50)

Carnassial teeth (p. 51)

Lower jaw of *Machairodus*

Giant sloth, more than 4 m (12 ft) tall (Pleistocene)

THE MAMMALS CONTINUED to evolve and diversify (change) and during the Miocene and Pliocene periods they became more "modern" looking. In Asia, North America, and Europe, more than three-quarters of Pliocene mammal species belonged to groups in existence today. In Australia and South America, land masses isolated for millions of years by continental drift, there were numerous marsupial (p. 30) mammals. Two million years ago South America became joined to North America, and more placental (p. 34) mammals from the north spread south. Australia is still physically isolated, and still has a wider range of marsupials than South America.

CONSPICUOUS CANINES
This lower jaw is from a Miocene sabre-toothed cat (the "sabre tooth" was in the upper jaw). Well-developed muscle-anchorage points in the face and neck region indicate it opened its mouth wide and stabbed its prey to death.
Species shown: *Machairodus aphanistus* (Greece)

ICE-AGE RHINO
An upper molar tooth from a woolly rhino of the Pleistocene shows how the folds of enamel and dentine (p. 50) were ground flat by chewing.
Species shown: *Coelodonta antiquitatis* (Devon, UK)

Well-developed pattern

Upper jaw of *Dorudon*

WHALE'S BONE
In the water, as on the land, new mammal species were evolving while others died out. This is the upper jaw of an extinct Eocene whale showing serrated teeth for gripping slippery prey.
Species shown: *Dorudon osiris* (Egypt)

ANCIENT GIRAFFE *below*
Sivatherium was a Pleistocene relative of the giraffe, although with shorter legs and neck and longer horns than today's version.
Species shown: *Sivatherium maurusium* (Tanzania)

Serrated teeth

A KNUCKLE-WALKER?
This is the "toenail" bone from *Chalicotherium*, a strange, extinct Miocene mammal related to rhinos and horses. Its front limbs were much longer than its back ones, and it may have "knuckle-walked" like a gorilla.
Species shown: *Chalicotherium rusingense* (Kenya)

Plesiaddax skull from the side

Reconstruction of *Sivatherium*, showing antlers behind bony forehead "cones"

Antler of *Sivatherium*

UNGULATE SKULL
Many new kinds of ungulates (hoofed mammals) came during the Miocene, especially horned ones. *Plesiaddax* was a type of antelope related to the musk ox of today.
Species shown: Plesiaddax depereti (China)

CAINOZOIC ERA (Tertiary)					Quaternary	
66	55	37.5	24	5	1.7	0.01
Palaeocene period Mammals rapidly diversify, but are still unlike those alive today	**Eocene period** The first primates and bats appear. Early horses appear	**Oligocene period** The first mastodons appear, and many relatives of the rhino	**Miocene period** Apes present. More modern plant-feeding mammals become abundant	**Pliocene period** The first humans evolve	**Pleistocene period** Ice-Age mammals abundant as the ice caps advance and retreat	**Holocene period** Modern mammals. Humans increase on all continents

RECENTLY EXTINCT CAVE-DWELLER
The cave bear was larger than any bear of today, and it was around with the early humans, as the scene below depicts. Some of its remains come from caves, especially in the Pyrenees and Alps of Europe.
Species shown:
Ursus spelaeus (Germany)

Canine teeth for stabbing prey

Cave bear skull from the side

Molar teeth for crushing meat

CAVE TAKE-OVER
This rather fanciful Pleistocene scene nevertheless shows some of the mammals with which our ancestors shared the countryside.

Mammal senses

THE CAT'S WHISKERS
Or in this case, the mouse's whiskers!
Whiskers are longer-than-normal hairs with
sensory cells embedded in the skin to detect
any movement. Most whiskers are on the face,
but some mammals have them on their legs,
feet, or back.

ONE REASON for the mammals'
success is their "good sense" - their
generally well-developed senses of
sight, hearing, smell, taste, and touch.
Each sense has been moulded by evolution to fit its owner's way of life. Good
vision would be of little use to an underground mammal like a mole (p. 56),
so this creature has poor eyesight; but
it has an extremely sensitive muzzle that
combines touch and smell in order to find
food (chiefly earthworms that emerge from
the walls of its burrow). We humans depend
on our sense of vision. It is estimated that four-fifths of what the human brain
"knows" enters via the eyes. So it is difficult for us to imagine the accuracy with
which a mammal with a good nose "smells" the world through scents and odours,
or how a bat "hears" its surroundings by echoed squeaks (p. 19). Yet, even though
we depend on our eyes so much, our vision is not tremendous -
other mammals, such as some species of squirrels, have
much sharper sight. On the plus side, however, the
primates (including humans and bushbabies)
are the main mammalian group with
colour vision. Most mammals see the
world in black-and-white.

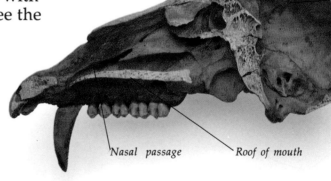

Brain cavity

Long, bushy tail

Nasal passage   Roof of mouth

SKULLFUL OF SENSES
This cutaway view of a baboon's skull shows how
a mammal's main senses are concentrated in the
head. Bony cavities protect the brain, the eyes,
the organs of smell, and the tasting tongue. The
mammalian brain is large in proportion to the
body, as it has to make sense of the mass of
information sent from the rest of the body.

PRICKING UP THE EARS
Many mammals,
including dogs, have a
good sense of hearing
and can move their
ears in the direction of
a sound. This gives
greater accuracy in
pinpointing where the
sound is coming from.

HUNTING BY TOUCH
The platypus grubs about in rivers
and streams for food, and finds its
prey of water worms, insects, and
"yabbies" (crayfish) almost entirely
by touch, as its bill is extremely
sensitive.

THE WORLD IN SMELLS
A pig with a "trained nose" snuffles for truffles
- underground fungi that will be dug up by its
owner and sold as an expensive delicacy.

FOOD-TESTING TONGUE
This lion tests its food partly by
smell and partly by taste. But
tongues not only taste, they do
other jobs. A mammal "licking its
lips" is cleaning its face (pp. 44-7).

FACING THE FRONT

The bushbaby seems all eyes and ears. This shy, nocturnal (active at night) member of the Order Primates (p. 8) has enormous eyes which can see in the darkest forest night, for locating prey and for leaping from branch to branch to avoid predators. Bushbabies are especially good at tracking small flying insects with their large swivelling ears; then, clinging with the back feet, they extend body and arms to snatch the insect as it flies past. The origin of this mammal's name may be from one of its calls, which is remarkably like the cry of a human baby - or it may be that the animal's appealing features look like those of a baby.

Sensitive ears pick up sound of flying insects

Huge eyes for accurately judging distances in the dark

Senegal bushbaby

EYES AT NIGHT

This odd-eyed white cat has one yellow eye and one blue eye. When photographed at night using flash its yellow eye reflects green from a shiny layer (tapetum) in the back of the eye. The blue eye lacks a tapetum so the blood vessels at the back of the eye glow red.

Rounded finger- and toetips for clinging to branches

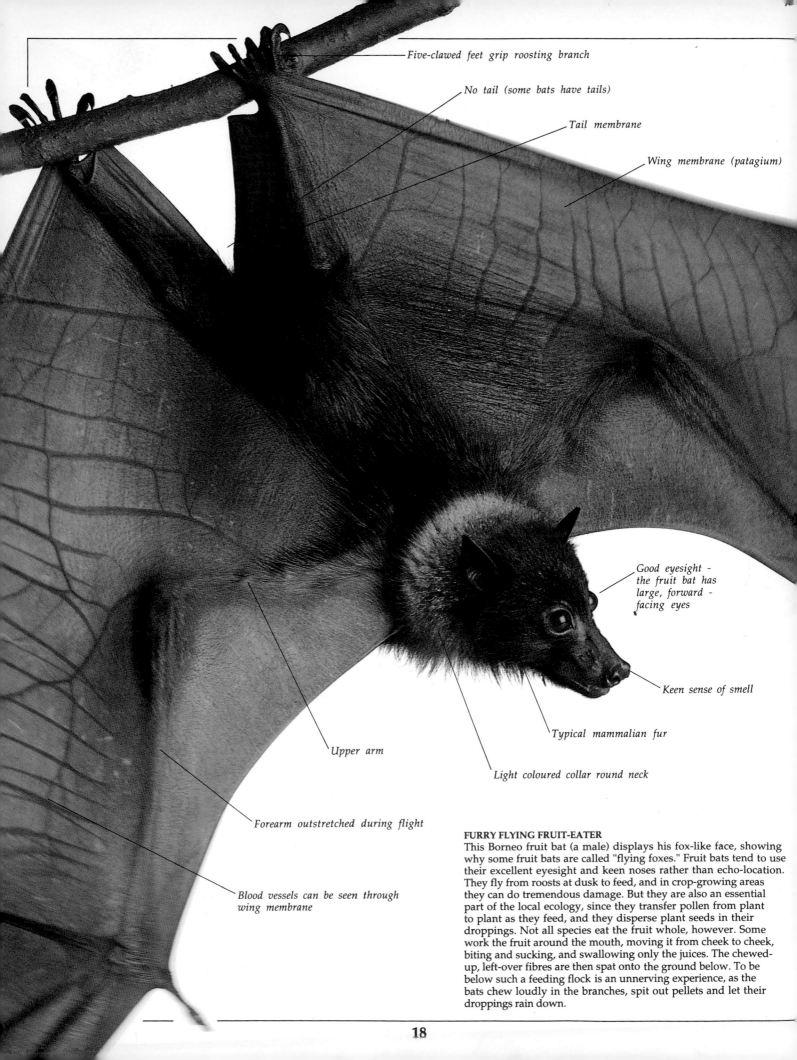

Five-clawed feet grip roosting branch

No tail (some bats have tails)

Tail membrane

Wing membrane (patagium)

Good eyesight -
the fruit bat has
large, forward -
facing eyes

Keen sense of smell

Typical mammalian fur

Light coloured collar round neck

Upper arm

Forearm outstretched during flight

Blood vessels can be seen through
wing membrane

FURRY FLYING FRUIT-EATER
This Borneo fruit bat (a male) displays his fox-like face, showing
why some fruit bats are called "flying foxes." Fruit bats tend to use
their excellent eyesight and keen noses rather than echo-location.
They fly from roosts at dusk to feed, and in crop-growing areas
they can do tremendous damage. But they are also an essential
part of the local ecology, since they transfer pollen from plant
to plant as they feed, and they disperse plant seeds in their
droppings. Not all species eat the fruit whole, however. Some
work the fruit around the mouth, moving it from cheek to cheek,
biting and sucking, and swallowing only the juices. The chewed-
up, left-over fibres are then spat onto the ground below. To be
below such a feeding flock is an unnerving experience, as the
bats chew loudly in the branches, spit out pellets and let their
droppings rain down.

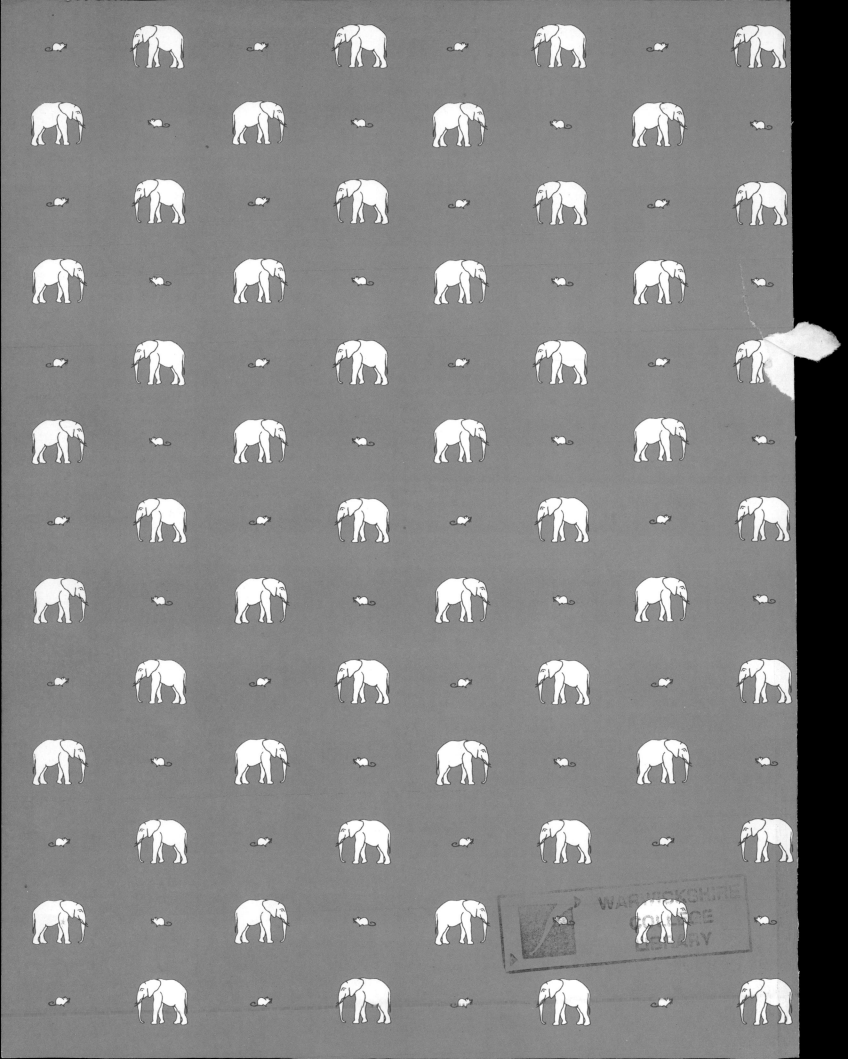

Flying mammals

MANY MAMMALS can leap and bound. Some can swim and dive. But only the bats can fly. Bats are the second most numerous group of mammals, in terms of species (p. 9). They vary enormously in size, from the tiny hog-nosed bat with a wingspan of 14 cm (5 in), to the sizeable flying foxes with a body the size of a small dog and outstretched wings 2 m (6 ft) across. The bat's flapping wings, unique among mammals, are made of thin sheets of muscle and elastic fibres covered by skin. The bones of the arm and second to fifth fingers support the wing; the "thumb" (first finger) is like a claw and used for crawling about, grooming, and, in some species, fighting and holding food. The muscles that power the wings are the ones you use to "flap" your arms, but proportionally many times stronger. Some bats can fly at more than 50 kph (30 mph). Bats are also among the most sociable of mammals. They roost together in their thousands in a cave or other suitable site. Some species cooperate in the nightly search for food. Males and females call to each other during the breeding season, and baby bats jammed like pink jelly-babies into nursery roosts squeak loudly as the mothers return from hunting.

Pegasus - the legendary flying horse

5th finger
4th finger
2nd finger
3rd finger
1st finger (clawed "thumb")

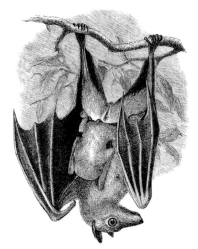

GLIDERS *right*
Bats are the only mammals capable of flight, but other mammals such as the marsupial gliders and colugos glide on the air using a membrane more like a parachute than a flapping wing.

BAT BABIES *left*
Clinging on to their mother's furry abdomen, bat babies suckle milk, just like other mammals.

FROM MOTHS TO BUDS TO BLOOD
Most bats are insectivores, eating moths, midges, flies, and other night-time flying creatures. The fruit bat (shown here) feeds on fruit, buds, and soft plant parts. The vampire bat feeds on blood.

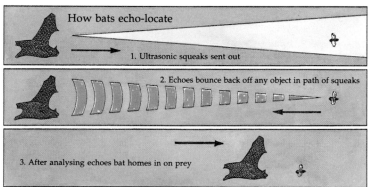

How bats echo-locate
1. Ultrasonic squeaks sent out
2. Echoes bounce back off any object in path of squeaks
3. After analysing echoes bat homes in on prey

"SEEING" WITH SOUND
In fact bats hear in the dark, using echo-location. They send out high-pitched squeaks through the mouth (1). The sound waves bounce off anything in their path and return to the bat's ears as echoes (2). The bat's brain computes the pattern of echoes, forming a "sound picture." The bat then homes in on the moth (3).

A VARIETY OF FACES
Among mammals, bats have some of the most interesting faces.

Horse-shoe bat   Leaf-nosed bat   Bat with fimbriated tongue

Furry coats

Fur, whiskers, wool, prickles, spines, and even certain horns - these are all made from hair, one of the mammal's trademarks. The advantages of a furry coat have helped the mammal group to success. Most important is the ability of hair to trap air and keep out cold and heat, wind and rain, and so insulate the mammal's body from the surroundings. Hairs grow from tiny pits in the skin, called follicles. They consist of cells cemented together and toughened with keratin, the same fibrous protein substance that strengthens the skin. Not all mammals have hair. Some, like the whales, lost it during evolution.

All the furs shown here come from museum collections - no animals were killed for this book

Hairseal fur

Mottled markings in fur

WOOL FOR WEARING
Sheep have been bred for their fur - wool - for centuries. Sheep's wool is a good insulator, absorbent yet springy, and takes up coloured dyes well. More than half the world's wool comes from the Southern hemisphere; three-quarters of it is used in the Northern hemisphere.

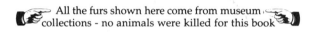

Spun and dyed wool, ready for weaving or knitting

Freshly sheared wool contains lanolin ("wool grease") used in cosmetics

WATER-TIGHT SEAL
The seal's skin contains many sebaceous (grease) glands that make its fur oily and water repellent. (Under the fur is a layer of blubber, like that of the whale.) Native peoples such as the Inuit (below) hunt seals for meat and also for their "seal-skin," made into boots and garments.

Inuit wearing seal fur hood

KILLED FOR BLUBBER
Whales do not have fur to insulate them from the cold ocean water. This job is done by blubber, a layer of fatty tissue under the skin, that also streamlines the whale's shape for efficient swimming. In some whales the blubber is 50 cm (20 in) thick. Processed blubber, or "whale oil," lit millions of lamps in days gone by, and was also used for making lubricants, soaps, cosmetics, margarines, and paints.

Section to show whale blubber

— Epidermis

— Dermis

— Blubber

— Connective tissue

— Fascia

— Muscles

FURRY SNOW
The Arctic fox grows a beautiful all-white winter coat to camouflage it in its snowy landscape. Another colour form, the blue, is greyish to brown in winter.

Polar colour form
Arctic fox fur

HAIRS ON GUARD
The opossum of North America has straight fur, unlike its marsupial cousin from Australia (right). The long, light guard hairs can be clearly seen projecting from the deep underfur.

Opossum fur

— *Underfur*

— *Guard hairs*

POSSUM'S PELT
The Australian possums tend to have crinkly or "crimped" hairs. The common brush-tailed possum is a cat-sized, tree-dwelling species. One of its local names is "silver-grey," on account of its fur.

Possum fur

— *Crimped hairs*

Karakul fur

Short, textured fur

SHORT AND DENSE
The beautiful, velvet-like pelts of young Karakul sheep are known as "Persian lamb." Sheep were domesticated some 10-12,000 years ago, and today there are about 350 breeds, raised for meat as well as wool.

Colobus monkey fur

Long, thick hairs

LONG AND LUXURIANT
Some species of colobus monkey became rare due to hunting for their fur, with its long, silky-looking hairs. The satanic black colobus has all-over, shiny black fur. Sadly, uninformed tourists still buy decorations and rugs made from colobus coats.

Beaver fur

Long guard hairs

KILLED FOR PELTS
Like many mammals, the beaver has two types of fur. One is a dense covering of short brown hairs, the underfur. The other is a sparser growth of longer, thicker hairs called the guard coat. The guard coat gives protection and camouflage, while the underfur supplies insulation and waterproofing. Hunting North American beavers for their pelts (furred skins) was so lucrative that wars were fought over land ownership, and the beaver fur trade helped open up much of North America in the 1700s and 1800s.

WATERPROOF FUR
Despite spending large amounts of its time under water, the water vole does not get "wet" as its long guard hairs keep the underfur dry.

NOT "NAKED"
We may think of ourselves as "naked apes," but we have lots of hairs. They are small and inconspicuous on the body, yet human head hair is typically mammalian. Fashions through the ages have sometimes emphasized "hairiness," such as the wig (left) worn by this learned judge.

Lock of human hair

WARNING STRIPES
The skunk's distinctive stripes stand out as a warning sign. If a predator tries to molest the skunk it carries out its threat display, raising its tail and stamping. Should the molester ignore the warning, the skunk may turn and spray it with foul liquid from the two anal glands.

BEAUTIFUL BELLY
Most spotted cats have spotted bellies. Lynx frequently have spotted backs as well.

Distinctive black and white stripes

Skunk fur

Lynx fur

Spotted belly fur

Hiding in the open

SMALL PLANT-EATING MAMMALS are very vulnerable when feeding out in the open. They have few ways of defending them-selves from their many enemies. Camouflage - resembling and "blending in" with the surroundings - helps them to remain undetected as long as they stay still. Predators, too, employ camouflage in order to sneak up unseen on their prey. Fur is well suited to this purpose. By variation in its length and in the cells that make the various pigments that colour it, virtually any shade and patterning can be achieved (p. 20).

PEBBLE WITH WHISKERS
Small rodents such as mice and voles are among the most vulnerable of all mammals. Their main defences are sharp senses and a quick dive into a nearby burrow, or good camouflage if stranded out in the open. This Arabian spiny mouse's fur blends with the dry sand, light-coloured pebbles, and parched wood of its semi-desert home.

BUILT-IN CAMOUFLAGE
Algae (tiny plants) grow on the coat of the slow-moving South American two-toed sloth. Its long outer guard hairs (p. 20) have grooves on them in which the algae grow. When the sloth is still (which it often is) in the dim forest light it merges with the foliage.

Dried wood

Light-coloured pebbles

Sand

LEAF WITH A TAIL
Meadow voles live in a variety of places, from grassland to woods and streambanks. They forage on the ground, which is usually littered with dark, dead or dying leaves and other plant parts. Voles are very busy animals, being active more or less round the clock, so visual camouflage is very important. A vole alerted by the soft wing-beat of a hunting bird might "freeze" in its tracks. In the dim light of dawn or dusk, or in the dappled shade under trees, it would be difficult to spot from above - as shown by this owl's-eye view.

Leaves are from a deciduous woodland

Dead leaves

Damp wood

HIDING FROM THE ENEMY
Humans at war imitate nature and camouflage themselves, their vehicles, and weapons. Standard combat clothes are "natural" greens and browns, mottled and patched to break up the soldier's outline in woods and scrubland. Snow-country outfits must be warm and white, like the Arctic fox's winter coat (p. 20).

DISGUISING THE OUTLINE
The Malayan tapir's striking color-ation of white back and belly but black everything else is a fine example of disruptive coloration. In the dark night-time forest the pattern breaks up the tapir's bulky body outline, making its distinctive shape less recognizable to predators. The young tapir is dappled white, a similarly disruptive device.

A spiny coat

Up to 5,000 sharp, stiff spines, sticking out at all angles, are enough to put off most predators. The spiny coat is the main defence of the Western, or European, hedgehog, a familiar mammal in the gardens, hedgerows, parks, and woodlands of Europe. Each spine is a hair modified during evolution into a sharp, stiff spike about 2-3 cm (1 in) long. The hedgehog's behaviour has evolved in tandem with its spines, so that when in serious trouble it rolls into a ball shape and waits for danger to pass.

As danger passes the head and front legs emerge

Hedgehog cautiously begins to unroll

Fully rolled hedgehog has no vulnerable parts

3 ALL-CLEAR
The hedgehog has decided that the main threat is over and now is the time to leave. Its head straightens and is first to protrude from the ball, so that the animal can smell, hear, and see clearly. Also beginning to emerge are its front legs. The hedgehog has surprisingly long legs, usually obscured under its mantle of spines. It can run well, burrow, clamber over low walls, and swim when it needs to.

2 CAUTIOUS PEEP
The spines physically intimidate the enemy, and they also act as a springy cushion should the hedgehog be pushed down a slope or against a tree. After a few moments of calm, the hedgehog relaxes slightly and peeks out of its prickly protection. Its eyesight is relatively poor, but the sense of smell is keen, and vibrations in the ground made by a creature moving nearby are felt via the spines.

1 ALL-OVER PROTECTION
In the face of danger, the hedgehog quickly tucks in its head, legs, and tail, and arches its back into a U-shape. A "cloak" of muscle under the loose skin pulls itself down over the head, sides, and rear. A band of muscle running around the edge of this cloak contracts, acting like a drawstring to pull the mantle of spines together around the underparts. The spines are automatically erected in the process. This defensive behaviour produces the tight ball that presents nothing but spines to the molester.

DEADLY ENEMY
The fox hunts many smaller mammals, including hedgehogs. It may poke and prod at a tightly rolled hedgehog for some time, in an attempt to make the animal uncurl and run off, whereupon the fox claws at the vulnerable belly.

Feet barely visible

Head remains tucked under

STRANGE BEHAVIOUR
Hedgehogs have often been seen to chew something foul (such as the dead toad, left), and then flick and spit their frothy saliva over their spines. It is not clear why. One theory is that this "self-anointing" is part of the animal's defence, helping to deter predators.

BABY'S DEFENCE
A baby hedgehog's first coat of rubbery spines lies flat under its skin at birth, but pops up within a few hours. The baby cannot roll up until it is 11 days old. Its main defence is to jerk its head upwards, stabbing predators on the nose.

4 FLIP-OVER
Should the hedgehog continue to unroll while lying on its back, its vulnerable underparts would be exposed to any predators. To prevent attack, the hedgehog executes a quick flip-over manoeuvre to land on its belly, keeping its feet tucked in and its head well down for continued protection.

5 PREPARING TO MOVE OFF
If there is no sign of renewed threat the hedgehog uncurls further. Its head emerges to reveal which end is which, and, sniffing and with whiskers quivering, it looks about for a suitable refuge, preferably a dark tangle of brambles and undergrowth.

Head emerges to investigate surroundings

6 QUICK EXIT
Defence gives way to escape, and the hedgehog scurries off to safety. This animal can move surprisingly fast when at risk - at about the speed of a human's quick walk - with its body held off the ground. But when foraging peacefully for slugs, worms, insects, and fallen fruit, it shuffles flatly among leaves and vegetation.

HEDGEHOG RELATIVE?
The echidna of Australia and New Guinea has a coat of defensive spines similar to the hedgehog. Yet it is only distantly related, having evolved the same system of defence separately. The hedgehog gives birth to live young, while the echidna lays eggs (p. 31).

Hedgehog moves at speed to safer place

Designs for defence

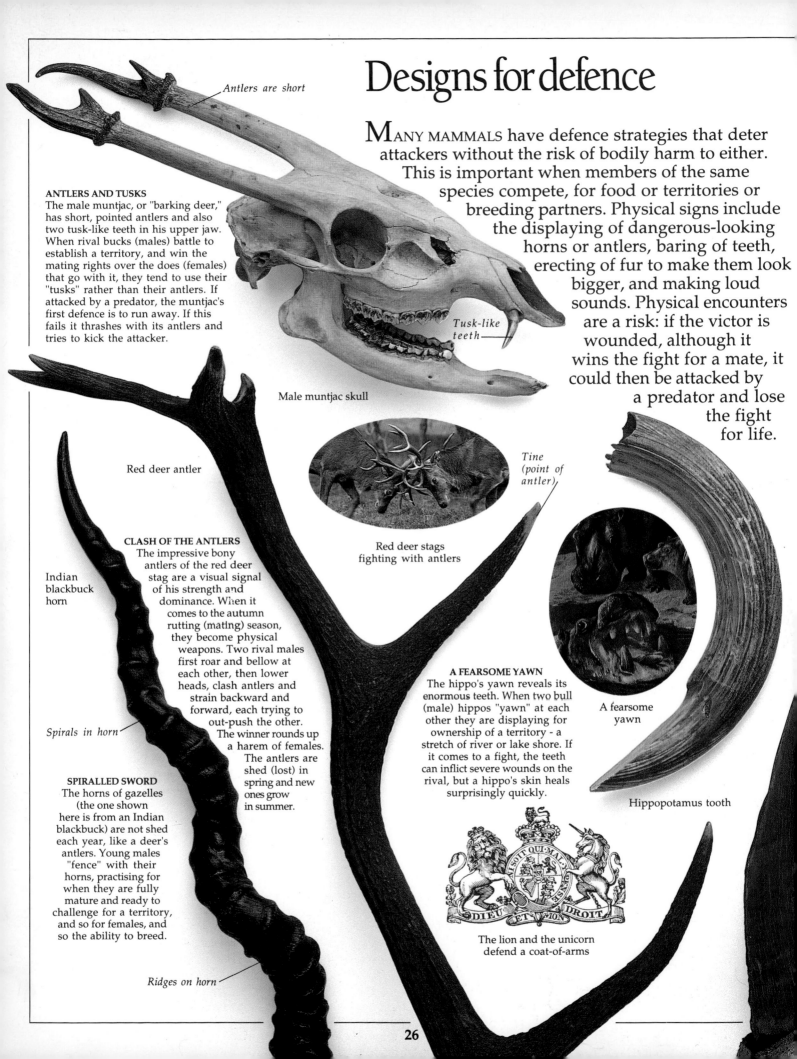

ANY MAMMALS have defence strategies that deter attackers without the risk of bodily harm to either. This is important when members of the same species compete, for food or territories or breeding partners. Physical signs include the displaying of dangerous-looking horns or antlers, baring of teeth, erecting of fur to make them look bigger, and making loud sounds. Physical encounters are a risk: if the victor is wounded, although it wins the fight for a mate, it could then be attacked by a predator and lose the fight for life.

Antlers are short

ANTLERS AND TUSKS
The male muntjac, or "barking deer," has short, pointed antlers and also two tusk-like teeth in his upper jaw. When rival bucks (males) battle to establish a territory, and win the mating rights over the does (females) that go with it, they tend to use their "tusks" rather than their antlers. If attacked by a predator, the muntjac's first defence is to run away. If this fails it thrashes with its antlers and tries to kick the attacker.

Tusk-like teeth

Male muntjac skull

Red deer antler

Tine (point of antler)

Indian blackbuck horn

CLASH OF THE ANTLERS
The impressive bony antlers of the red deer stag are a visual signal of his strength and dominance. When it comes to the autumn rutting (mating) season, they become physical weapons. Two rival males first roar and bellow at each other, then lower heads, clash antlers and strain backward and forward, each trying to out-push the other. The winner rounds up a harem of females. The antlers are shed (lost) in spring and new ones grow in summer.

Red deer stags fighting with antlers

A FEARSOME YAWN
The hippo's yawn reveals its enormous teeth. When two bull (male) hippos "yawn" at each other they are displaying for ownership of a territory - a stretch of river or lake shore. If it comes to a fight, the teeth can inflict severe wounds on the rival, but a hippo's skin heals surprisingly quickly.

A fearsome yawn

Spirals in horn

SPIRALLED SWORD
The horns of gazelles (the one shown here is from an Indian blackbuck) are not shed each year, like a deer's antlers. Young males "fence" with their horns, practising for when they are fully mature and ready to challenge for a territory, and so for females, and so the ability to breed.

Hippopotamus tooth

The lion and the unicorn defend a coat-of-arms

Ridges on horn

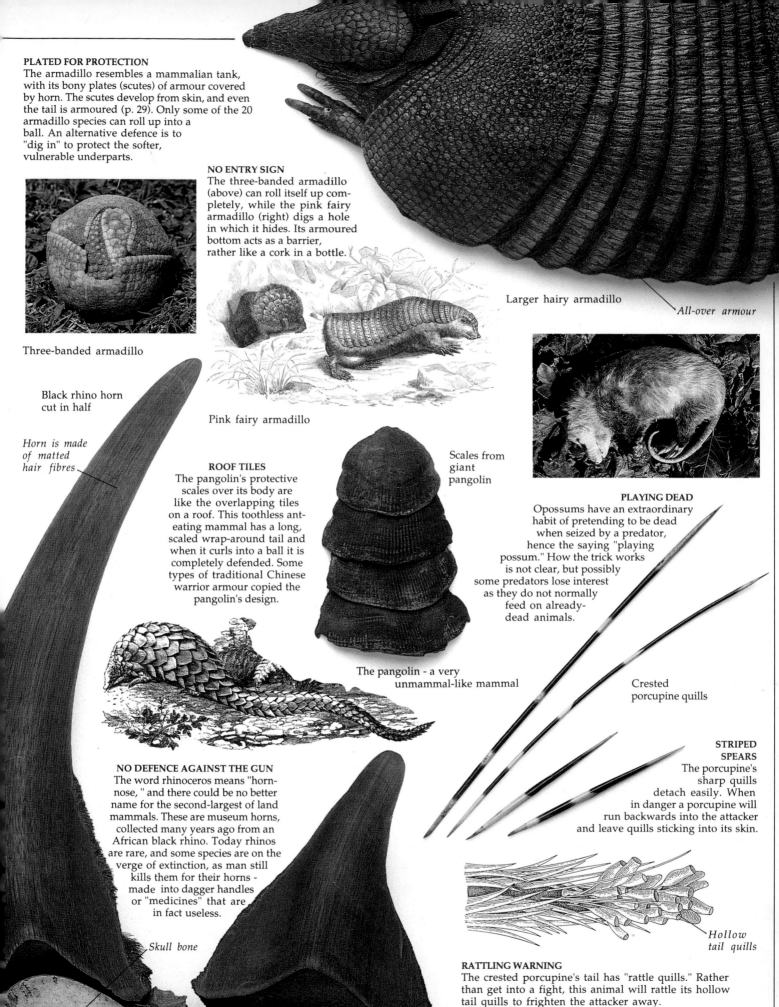

PLATED FOR PROTECTION

The armadillo resembles a mammalian tank, with its bony plates (scutes) of armour covered by horn. The scutes develop from skin, and even the tail is armoured (p. 29). Only some of the 20 armadillo species can roll up into a ball. An alternative defence is to "dig in" to protect the softer, vulnerable underparts.

NO ENTRY SIGN

The three-banded armadillo (above) can roll itself up completely, while the pink fairy armadillo (right) digs a hole in which it hides. Its armoured bottom acts as a barrier, rather like a cork in a bottle.

Larger hairy armadillo

All-over armour

Three-banded armadillo

Black rhino horn cut in half

Horn is made of matted hair fibres

Pink fairy armadillo

ROOF TILES

The pangolin's protective scales over its body are like the overlapping tiles on a roof. This toothless ant-eating mammal has a long, scaled wrap-around tail and when it curls into a ball it is completely defended. Some types of traditional Chinese warrior armour copied the pangolin's design.

Scales from giant pangolin

PLAYING DEAD

Opossums have an extraordinary habit of pretending to be dead when seized by a predator, hence the saying "playing possum." How the trick works is not clear, but possibly some predators lose interest as they do not normally feed on already-dead animals.

The pangolin - a very unmammal-like mammal

Crested porcupine quills

STRIPED SPEARS

The porcupine's sharp quills detach easily. When in danger a porcupine will run backwards into the attacker and leave quills sticking into its skin.

NO DEFENCE AGAINST THE GUN

The word rhinoceros means "horn-nose, " and there could be no better name for the second-largest of land mammals. These are museum horns, collected many years ago from an African black rhino. Today rhinos are rare, and some species are on the verge of extinction, as man still kills them for their horns - made into dagger handles or "medicines" that are in fact useless.

Skull bone

Hollow tail quills

RATTLING WARNING

The crested porcupine's tail has "rattle quills." Rather than get into a fight, this animal will rattle its hollow tail quills to frighten the attacker away.

Bones from horse's tail

What is a tail for?

Elephant tail

ON THE INSIDE, the mammalian tail is a continuation of the spine, made up of a column of vertebrae (backbones). But on the outside, tails are as varied in size, shape, and function as their owners. They can be fluffy "scarves" for winter warmth, whisks or switches to get rid of flies, or strikingly patterned "flags" that convey a mammal's mood and intentions to others. Most of us are familiar with a dog wagging its tail when happy and lowering it between its legs when scolded, or a cat twitching its tail when annoyed. "Tail talk" is common among mammals, with small variations in the posture and movement of the tail communicating aggression, submission, and other behaviours. Relatively few mammals lack tails. They include, of course, ourselves. The evolutionary remnant of our "tail" is a small knob of four or five fused vertebrae, the coccyx, at the base of the human spine.

Horse tail

HAIRY FLY-WHISK
The horse's tail is made of hundreds of long, thickish hairs that the animal uses as a fly-whisk to swish away irritating pests. The last 15 or so vertebrae of the spine occupy about half the tail's length (above), and these are moved by lengthwise muscles. Holding the tail up is a sign of arousal (as when courting), while tail-lashing indicates the horse is angry, irritated, or perhaps in pain.

TRUNK TO TAIL
The largest living land animal has thick, sparsely haired skin - but a comb of wiry hairs at the end of its tail. When elephants walk in single file, each may curl its trunk around the tail of the one in front.

Thick, stiff hairs

Tail is made from long hairs

Fallow deer tail tuft

TAIL TUFTS
The lion's long, mobile tail has a small tuft of dark hairs at its tip (far left). Young cubs often play with the tail tuft of an adult, practising pouncing. The fallow deer's tail (left) is dark on top with a white under-side. The fur on the body under the tail is also whitish, with black stripes. When danger threatens the tail is held erect and "flashes" a visual warning to other deer in the herd.

Lion tail tuft

28

ROUGH FOR GOOD GRIP
Possums are marsupial tree-dwellers from Australia and Southeast Asia. This tail of a New Guinea possum shows the bare scaly skin on the underside at the tip. The rough skin gives a better grip than furry skin. The possum's tail is prehensile, able to wind round branches and function as a "fifth limb" (see below).

HANGING AROUND
The spider monkey has a muscular prehensile tail - a good safety feature for life spent in the trees of South American rain forests.

Ring-tailed lemur tail

New Guinea possum tail

Canadian beaver tail

RUDDER AND ALARM
The Canadian beaver's flat, scaly tail is used as a rudder when the animal swims with its broad, webbed back feet. The tail can also be flapped up and down to give extra thrust in an emergency. And if the beaver is alarmed, it slaps its tail down on the water surface with a loud "smack" to warn its companions.

Scaly skin for good grip

Large scales on tail

BUSHY BRUSH
The furry "brush" of the red fox makes an excellent wrap-around warmer to keep the animal snug during winter. It used to be thought that foxes were solitary hunters. Now it is known that they are social animals, and their tails give visual signals to others in the family group. The tip or "tag" may be dark or white.

Red fox tail

Position of scent gland - this has a role in social communication

RAT-TAILED
The black-tailed tree rat has the unfurry, scaly tail typical of rats and mice. The tail is used as an aid to balance.

Black-tailed tree rat tail

White tag

Tail has no fur

BLACK TIP
The stoat's winter white coat (brown in summer) is good camouflage in the snow. Yet its tail-tip remains black. It is now thought that this may confuse predatory birds such as owls, which will dive at the black tip, rather than the vulnerable head of the stoat.

PLATED FOR PROTECTION
Even the armadillo's tail is well armoured, like the rest of its upper body surfaces (p. 27). The tough, horny plates develop from hardened skin.

Armadillo tail

Horny plates

SMELLY FLAG
Ring-tailed lemurs are social mammals, active by day and spending less time in the trees than other species in the lemur group. As they walk on all fours, their distinctive banded tails are held up in the air. When challenging other males for a place in the group's hierarchy, a lemur wipes its tail over scent glands on its shoulders and fore-arms and then flicks the tail over its head, spreading the scent into the air.

THE TAIL OF THE WHALE
The muscular tail is made of two large flaps or "flukes." The swimming power comes from the muscles in the back moving the flukes up and down.

Horsfield's flying squirrel tail

TAIL FOR FLYING
The flying squirrel swoops from tree to tree, parachuting on flaps of skin down the side of its body. The flattened tail acts as a rudder and an air-brake.

Flattened to help with steering

Banding acts as visual signal to other lemurs

An early birth

Australian marsupials

MOST MAMMALS DEVELOP in their mother's womb (uterus). When born they are well-formed, and in many species they are up and about within hours of birth (p. 35). Pouched mammals, or marsupials, don't follow this system. What sets them apart from all other mammals is the way they reproduce. The Eastern grey kangaroo is a typical example: its baby grows for only five weeks in the womb. When born it is just 2.5 cm (1 in) long, naked, blind, and unrecognizable as a kangaroo (see opposite). It wriggles from the birth opening (which is not the same as the birth canal of other mammals, p. 35) to the teat in the mother's pouch. As it takes the teat in its mouth and sucks, the teat swells and the baby is "stuck" there as it continues to grow and develop. The pouch, therefore, acts as a sort of "external womb" where the baby continues its development. After a while the baby's jaws enlarge and it can let go of the teat. Later it grows enough to leave the pouch for short periods. After about 10 months the youngster is too big to get into the pouch.

Adult female
red-necked wallaby

Four-month-old male
red-necked wallaby

KANGAROOS AND WALLABIES

This mother red-necked (or Bennett's) wallaby with her "joey" (baby) is a typical member of the kangaroo and wallaby family. There are about 50 species in the family, out of the 120 or so marsupial species found in Australia. There is no real difference between a kangaroo and a wallaby: larger species tend to be called kangaroos, and smaller ones wallabies. The scientific name for the family is Macropodidae, which means "big feet." This reflects the way they move - bounding along in great leaps on the huge feet, using the tail as a counterbalance. Some large kangaroos can travel at 60 kph (nearly 40 mph). When grazing on plants (all kangaroos and wallabies are herbivorous) they move slowly, resting their tails and front paws on the ground as they swing the back legs forward. At rest, they sit back on their tails or lie lazily in the shade of a tree. The red-necked wallaby was one of the first marsupials seen by Europeans, when the First Fleet anchored in Sydney Cove in 1788. Its traditional name is the "brusher," since it prefers brush and wooded areas rather than more open country. The four-month-old joey is now beginning to leave his mother. But at the first sign of danger he hops back to the safety of the pouch. He leaves the pouch at nine months, but will not be weaned (p. 33) until about 12 months old.

An even earlier birth

Only three of the 4,000 or so mammal species lay eggs. These are the platypus of Australia, the short-beaked echidna of Australia and New Guinea, and the long-beaked echidna of New Guinea. They are the only members of the Order Monotremata (p. 8), the "egg-laying mammals." When the white, leathery-shelled eggs hatch, about two weeks after laying, the young feed on mother's milk. The milk oozes from enlarged pores on to the skin, where the babies drink it; monotremes have no teats (p. 36).

Platypus head

Echidna egg

Echidna head

MARSUPIAL MONKEY?

Some opossums, marsupials of the Americas, look rather like monkeys - although they are not closely related, other than being mammals. This woolly opossum lives in the tropical forests of Central America and northern South America. Like a monkey it has large, forward-facing eyes, to judge distances accurately as it moves through the branches. It also has a prehensile tail, like some South American monkeys. It is a fruit- and nectar-feeder, like many monkeys. Yet its breeding is typically marsupial. After birth, the babies hold on to the teats in the pouch continually. As they grow, they are able to clamber about their mother and get a ride.

After its epic journey to the pouch, the developing kangaroo attaches itself to the teat and suckles milk (p. 36), just like any other mammal

LOST FOR EVER?

The thylacine, or "Tasmanian tiger," is a striped, wolf-like marsupial. Or rather it was, since it is probably extinct. The last captive thylacine died in 1936 in Hobart Zoo, Tasmania. From 1938 these creatures were protected by law, having been hunted for their raids on sheep and poultry. Possible sightings are reported now and again in Tasmania's wild hill country, and even on mainland Australia, but many naturalists believe this marsupial has gone forever.

Fast breeders

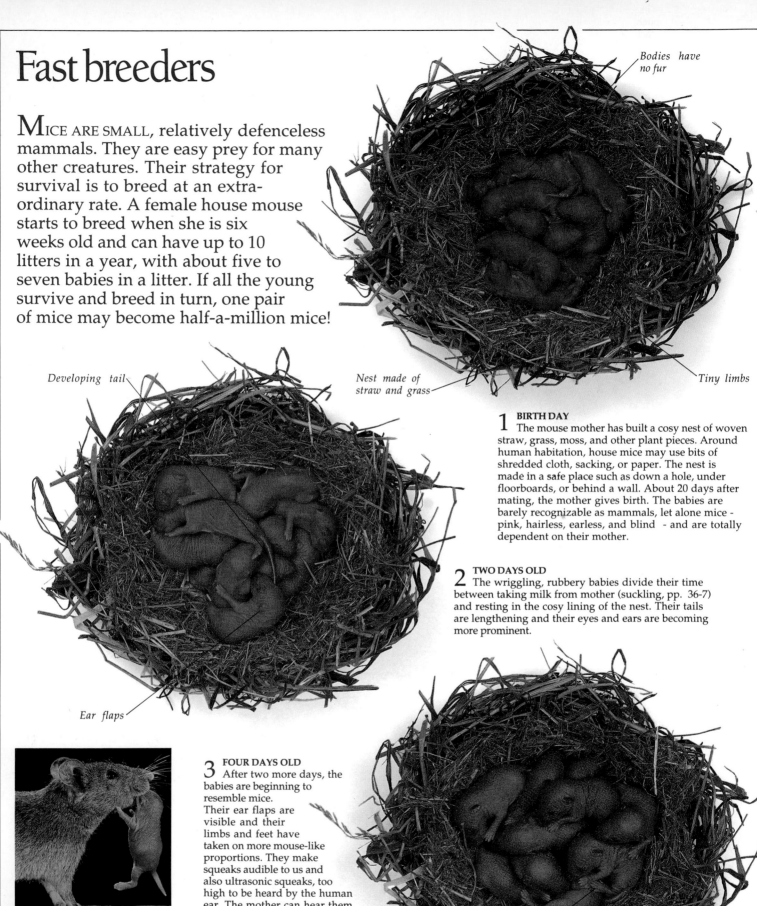

MICE ARE SMALL, relatively defenceless mammals. They are easy prey for many other creatures. Their strategy for survival is to breed at an extraordinary rate. A female house mouse starts to breed when she is six weeks old and can have up to 10 litters in a year, with about five to seven babies in a litter. If all the young survive and breed in turn, one pair of mice may become half-a-million mice!

Bodies have no fur

Tiny limbs

Nest made of straw and grass

Developing tail

Ear flaps

1 BIRTH DAY
The mouse mother has built a cosy nest of woven straw, grass, moss, and other plant pieces. Around human habitation, house mice may use bits of shredded cloth, sacking, or paper. The nest is made in a safe place such as down a hole, under floorboards, or behind a wall. About 20 days after mating, the mother gives birth. The babies are barely recognizable as mammals, let alone mice - pink, hairless, earless, and blind - and are totally dependent on their mother.

2 TWO DAYS OLD
The wriggling, rubbery babies divide their time between taking milk from mother (suckling, pp. 36-7) and resting in the cosy lining of the nest. Their tails are lengthening and their eyes and ears are becoming more prominent.

3 FOUR DAYS OLD
After two more days, the babies are beginning to resemble mice. Their ear flaps are visible and their limbs and feet have taken on more mouse-like proportions. They make squeaks audible to us and also ultrasonic squeaks, too high to be heard by the human ear. The mother can hear them, of course. If she leaves the nest for too long and the babies get cold, they will call to her ultrasonically, as if asking "Please come and warm us up!"

RETURN OF THE WANDERER
The mouse is a good mother. She locates babies who wriggle or fall from the nest, partly by following their squeaks, and will carry them by mouth back to safety.

Toes have developed

Fur starts to appear

SIGN OF THE MAMMAL
Feeding the young on milk is a unique mammalian trait, and the mother mouse feeds her babies regularly. This gives them the energy they need to grow at such an amazing rate.

Eyelids are open

4 SIX DAYS OLD
The youngsters take on a house-mouse brown as their fur appears. This is a risky phase, since their movements and their squeaks are becoming more forceful, and so their nest is in greater danger of being found by prowlers. The mother continues to suckle; her offspring will not be weaned (off milk and onto "solids" such as seeds and grains) until they are about 18 days old. The father is long gone. He takes little or no part in family life.

5 TEN DAYS OLD
The eyelids are now open and the young mice can see, although they are short-sighted creatures. They are also more mobile, with increasingly coordinated movements. In many mammals, young would have reached this stage of development in the womb, and they would now be born (see the kittens on pp. 34-5). However, mice rely on large litters to keep up their numbers, so the young are born in a "premature" state as the mother would be unable to carry so many big babies in her womb.

Body is covered with fur

6 FOURTEEN DAYS OLD
The mice are now becoming curious about their surroundings and leave the nest for short periods. In a few more days the young mice will be ready to make their own way in the world. Soon they will face hazards alone, such as predators, lack of food, exposure to the elements - and overcrowding as they themselves begin to breed.

At two weeks old mice start exploring away from the nest

Nest is now too small

Nine lives

A MAMMAL WHOSE BABIES DEVELOP INSIDE THE MOTHER'S WOMB is known as a placental mammal. The womb protects the babies until they are fairly well developed and a special organ called the placenta supplies the baby with food and oxygen. Cats are placental mammals and the babies are born with all their fur. Compare the kittens with the baby mice on p. 32 (also placental) and the baby wallaby on p. 30 (a marsupial). In many species, the gestation period (length of time it takes for the baby to develop in the womb) is linked to body size. In shrews it is about two weeks; in rhinos, 16 months. The birth itself is a dangerous time for mother and babies, since they are unable to flee and the odours of the birth fluids may give them away. Birth is generally a private affair, and even group-dwelling mammals such as deer, leave their companions for a safe spot in which to bear young.

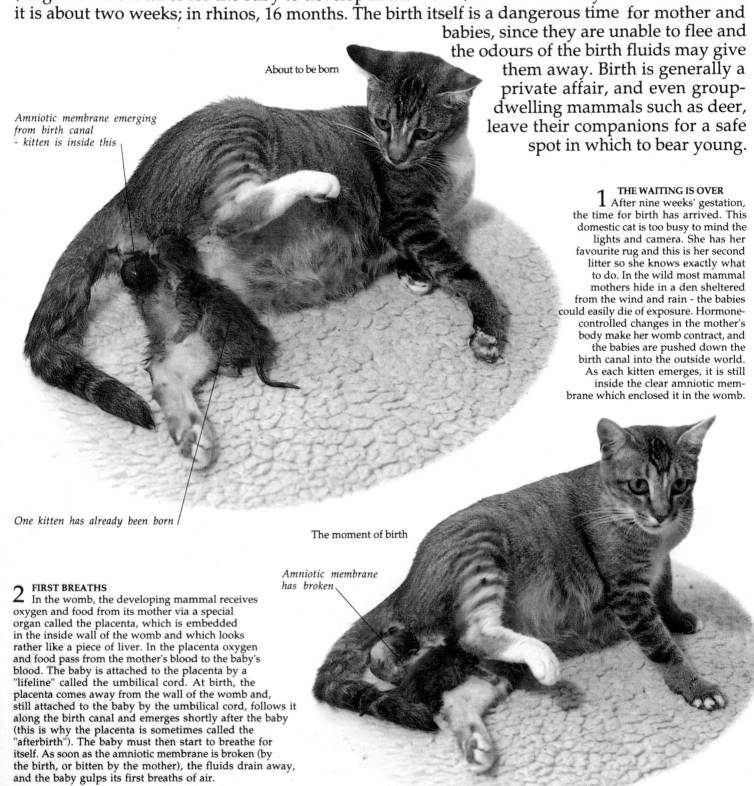

About to be born

Amniotic membrane emerging from birth canal - kitten is inside this

One kitten has already been born

The moment of birth

Amniotic membrane has broken

1 THE WAITING IS OVER
After nine weeks' gestation, the time for birth has arrived. This domestic cat is too busy to mind the lights and camera. She has her favourite rug and this is her second litter so she knows exactly what to do. In the wild most mammal mothers hide in a den sheltered from the wind and rain - the babies could easily die of exposure. Hormone-controlled changes in the mother's body make her womb contract, and the babies are pushed down the birth canal into the outside world. As each kitten emerges, it is still inside the clear amniotic membrane which enclosed it in the womb.

2 FIRST BREATHS
In the womb, the developing mammal receives oxygen and food from its mother via a special organ called the placenta, which is embedded in the inside wall of the womb and which looks rather like a piece of liver. In the placenta oxygen and food pass from the mother's blood to the baby's blood. The baby is attached to the placenta by a "lifeline" called the umbilical cord. At birth, the placenta comes away from the wall of the womb and, still attached to the baby by the umbilical cord, follows it along the birth canal and emerges shortly after the baby (this is why the placenta is sometimes called the "afterbirth"). The baby must then start to breathe for itself. As soon as the amniotic membrane is broken (by the birth, or bitten by the mother), the fluids drain away, and the baby gulps its first breaths of air.

3 CUTTING THE CORD

The kitten waits by the mother's tail until the placenta comes out. During the wait, the blood in the umbilical cord clots and prevents the kitten from bleeding to death when the mother chews through it. The mother will then eat the placenta as it is a good source of nutrients at a time when she cannot get out to feed. Also, its odour would attract unwelcome attention from predators and flies if she did not clean it up. After this the mother licks the kitten, drying its fur so that it fluffs up and keeps the baby warm. In the meantime, the first-born kitten has scrambled along its mother's body, using smell and touch to find her teats to suckle milk (p. 36). It is hard work for the mother - as the kittens are born within about half an hour of each other there is always one to be licked and cleaned. The mother wildcat ferociously attacks any animal that dares to interrupt her at this time. Even her kittens struggle and spit if threatened. The domesticated mother cat appreciates help and attention from a trusted human but her kittens hiss and bare their toothless gums like wild kittens if they smell someone near them.

Cutting the umbilical cord

Mother uses her teeth to cut the umbilical cord

First-born kitten is already feeding

A heap of brothers and sisters

4 IT'S HARD WORK BEING BORN

The damp newborns look squashed and fatigued after their birth. Their eyes and ears are sealed so that they are blind and deaf, but they are not as helpless as they might seem. In fact, they are very active and they are built to last. If the mother accidentally sits or treads on them, they squeak vigorously and let her know about it.

UP AND ABOUT
As a contrast to the defenceless kittens, a newborn calf is soon able to walk and run. Evolution has ensured that hunted animals, especially in open habitats, spend as little time as possible giving birth.

5 HAPPY FAMILY

The kittens have all been born. It was a large litter, but the births were quick and easy, with no problems for the mother. She continues to lick and dry all the kittens repeatedly. Soon she will be able to lie back and sleep while the kittens suckle contentedly, dry and warm against her belly. The riskiest time is over.

Unique to mammals

A mare has two teats between her back legs. She nudges her foal towards them; the foal then feeds on average some 4 times each hour

THE MAMMARY GLANDS, unique to the mammalian mother, form in the skin. They resemble specialized sweat glands and grow in two "milk lines" on each side of the abdomen. Cats and dogs have several glands and teats along each side; in hoofed animals they are near the hind legs. In primates (including humans) they are on the chest, a site that may be connected with adaptation to a tree-dwelling life and the consequent need to hold the babies with the fore-limbs. During pregnancy, the mammary glands increase in size under the influence of the female hormones oestrogen and pro-gesterone. Milk production is stimulated by another hormone, prolactin. After birth a hormone called oxytocin, from the pituitary gland (just under the brain), causes the gland to release its milk and encourages formation of more milk. Milk is the young mammal's complete food, providing even the water it needs.

THE NEAT TEAT
Unlike kittens, puppies usually feed from whichever teat they can find. The teat (or nipple) is a rubbery-textured lobe of tissue. It fits neatly inside the baby's mouth, to minimize loss of milk as the baby suckles. The teat also acts as a shut-off valve to prevent leakage of milk after feeding.

The mother cat (p. 34) with her third, much smaller, litter

CONTENTED CAT AND KITS
Within an hour of birth, a kitten is suckling (sucking milk from its mother's teat). Since there is usually about 20 minutes between the births of successive litter-mates, and there are four or five kittens in an average litter, the first-born will already be suckling when the later ones arrive. The tiny kitten, although unable to see or hear, can smell - and can feel with its whiskers, fur, nose, and feet. It moves to the milk supply by scrabbling with its feet, first locating the warmth of the mother's body, then working its way along until it finds a teat. It "kneads" the teat with its feet and face, to stimulate milk flow. After an initial free-for-all, each kitten tends to settle into a routine and suckle from its "own" teat. If there is a large litter, the young may feed in "shifts."

Teats run the length
of the mother's abdomen

This teat is not needed
for this small litter

MATERNAL MANATEE
The manatee, a marine mammal, has
teats situated just behind her front
flippers, near her "armpits." The
youngster feeds underwater,
lying by its mother's side in calm
water. Sometimes the mother
holds the baby with her
flipper to prevent it
floating about in water
currents, a bit like
a human mother
cradling her
infant.

New-born kittens feeding
on their mother's milk

Each kitten has
its own teat

THE WOLF TWINS
The legendary founders of ancient Rome, human twins
Romulus and Remus, were supposedly suckled by a she-
wolf until discovered and raised by shepherds. It is
unlikely that wolf's milk could provide
the nutrients required by humans.

SEARCHING FOR THE NIPPLE
A human baby, unlike many other
mammals, loses weight slightly
after being born but regains birth
weight by one week. In the "rooting
reflex," when the baby's cheek is
stroked it turns to that side,
searching for the nipple - a useful,
built-in behaviour for a newborn.

Growing up

COMPARED TO OTHER ANIMALS, mammal parents invest a lot of time and energy in their young. An insect may lay hundreds of eggs and leave them to their own devices. A sea urchin casts thousands of eggs into the water and has nothing more to do with them. Mammals adopt a different strategy. In general, they have only a few offspring, but they look after them well. The young are cleaned, fed, kept warm, protected, taught, and generally cared for until they are self-sufficient. The degree of parental care varies, however, within the mammal group. We are at one end of the spectrum: human parents spend many years raising their children. The mother tree shrew is probably the worst parent. She leaves the young in a nest after birth and returns only once every couple of days. The female cat looks after her kittens until they are weaned and old enough to feed themselves. Kittens grow rapidly, as these pictures show, getting the energy they need for growth from their mother's milk (p. 36). By nine weeks old the kittens have grown enough to leave their mother. Compare this with the wallaby (p. 30) and the mice (p. 32).

HELPLESS AT SIXTY-THREE DAYS OLD
While the young cat has become self-sufficient, a human baby of the same age is relatively helpless. One of its most rewarding behaviours is to smile, which encourages affection and handling (and so warmth) and strengthens the mother-baby bond. But it will be many more years until it is fully independent.

Newborn kitten

Eyes and ears are closed

Fur has dried

1 BIRTH DAY
Kittens have their fur at birth. But living in the watery environment of the womb makes the baby look waterlogged. The "water" is amniotic fluid. (p. 34). The mother licks her offspring thoroughly and the fur is soon dry and shiny. The kitten is relatively helpless: it cannot see or hear (the eyes and ears are closed), and it cannot lift its head. But it can feel and smell, and push itself along, so that it soon finds the mother's teat and begins to feed (p. 36).

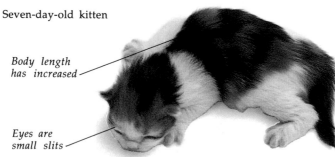

Seven-day-old kitten

Body length has increased

Eyes are small slits

2 SEVEN DAYS OLD
In a week, the kitten has doubled its weight from the 100 g (4 oz) or so at birth. Its eyes are just beginning to open. They cannot yet detect colours and shapes, these are just a jumble at first. The kitten must learn to recognize and make sense of what it sees, and this takes time. The mother cleans it and licks up urine and bowel motions. In the wild this is sensible behaviour, because a soiled and smelly nest would soon attract predators.

Twenty-one-day-old kitten

3 TWENTY-ONE DAYS OLD
By now the kitten's eyes and ears are functioning well, and it can hold its head up. Its weight has quadrupled since birth, and muscles are now stronger and more coordinated, and the legs are a bit longer, so that the youngster can just about creep along in a shuffling manner. If the kitten is in trouble it mews loudly, showing the first or "milk" teeth that appear between the age of two and three weeks.

Eyes fully open

Milk teeth

Legs are longer allowing creeping movement

Thirty-day-old kitten

4 THIRTY DAYS OLD

After feeding, the kitten's abdomen hangs down and only just clears the ground, since its legs are still relatively short. By now it is walking confidently and leaves the nest of its own accord to explore and play. One great change is the process of weaning: the kitten is starting to sample solid food and suckle less milk from the mother. She brings prey back to the nest and allows the kitten to examine it, so that it learns what it should hunt later.

Face is changing proportions and looks less babyish

Kitten can stand

Forty-two-day-old kitten

Head is still large in proportion to body

5 FORTY-TWO DAYS OLD

By six weeks of age, the kitten still has a large head and shortish legs, but its general body proportions are becoming more like those of an adult cat. It now leaves the nest for longer periods, exploring and playing with its litter-mates (domestic cats average about five kittens to the litter). Coordination is increasing and the kitten can run, jump, and climb, though it rarely ventures far from home. Solid food features more in the diet, but much nourishment still comes from the mother's milk. A female lemming of this age would already be pregnant with her first litter.

Sixty-three-day-old kitten

Tail is longer - more adult-like

6 SIXTY-THREE DAYS OLD

Most kittens are fully weaned by nine weeks. Although still part of a family, enjoying each other's company, they are now able to fend for themselves and could be separated from the mother. The young cat seems very playful but much of its activity has the important function of teaching it how to catch prey and avoid danger (p. 42). This tortoiseshell youngster, playing with red wool, is training its eye-paw coordination, sharpening its reactions and testing the grip of its claws - as well as finding out about the wool when this is caught.

The game of life...

Not a natural object, but the baby orang could be learning to manipulate a ripe fruit

IT IS DIFFICULT TO IMAGINE an ant or a leech playing. What we humans call play seems fairly restricted to mammals, with their well-developed senses and ability to learn and to be intelligent. Play occurs chiefly in young mammals. It is generally an un-serious business, seemingly carried out for its own sake, with none of the purpose found in adult behaviour patterns such as feeding or establishing a territory. Young chimps chase in a rough-and-tumble, badger cubs roll and frolic outside their sett, and even baby platypuses waddle around, squealing and yelping like puppies. Theories about why young mammals play are not in short supply. For the individual, it helps to develop strong muscles and good coordination. For survival, it trains a carnivore in hunting techniques or a herbivore in detection of and flight from threat. For social mammals it provides a basis in communication, in the use of sounds and body posture to convey messages such as dominance and submission that coordinate the group.

Testing the strength of the fabric

THE CHIMP AND THE CLOTH
This two-year-old male chimp was allowed to play with a strip of cloth. He had seen fabrics before, but this new piece deserved some general atten- tion to appreciate its colour and gauge its texture and strength (left) - and, as always, to test the faint possibility that it might be good to eat. Next came a series of actions to wear the fabric. The chimp watched his human companions closely while doing this. When the result got a response from them, usually laughter at the cloth becoming a "scarf" or "hat" (above) or a face-cover (far right), he was encouraged and experimented further. Later, he turned his attention to details and began studiously to unpick individual threads from the fabric (below). Many aspects of future behaviour are seen here, from arm- muscle strength needed to build a leafy "bed" each evening, to the fine dexterity of fingers required when grooming (p. 44) or feeding on small food items (p. 49).

Concentrating to unpick threads

Using the fabric as a hat

NO TERMITES HERE . . .
In the wild, chimps eat mainly fruits and leaves.
But they also take termites and ants from holes,
using sticks as tools (p. 49). Hole-examining is a
common behaviour and occurs in situations which
we might consider strange. Finding juicy termites
in the hole on the base of a toy building brick is
highly unlikely. But, for the chimp, you just
never know. . .

TOY TIME
We are so used to watching
human babies playing with
purpose-made toys that
we may lose sight of the
evolutionary origins of
such activity. In some
groups of people,
with less industrialized
ways, sticks and stones
and leaves make ideal
natural toys.

Using the fabric
as a face-cover

Arm exercises
improve strength
and coordination
for a life in the trees

...continues

P LANT-EATING MAMMALS may have to travel and search to locate food, but once found, the plant is relatively easy to "catch." For the hunting carnivore (meat-eater), finding food is more risky. It involves much effort, stalking or tracking prey. When it comes to the kill, there is a risk of injury from the victim's defences (p. 26). If the prey escapes, time and energy has been wasted. So it is not surprising that the play of young carnivores, such as cats and dogs, appears to mirror so many features of adult hunting life. To prevent misunderstanding and accidental injury, it is important that the animal wishing to play conveys this fact to its fellows, otherwise they might take its actions seriously. A puppy will make a "play-bow," crouching down on its chest, rump and back legs up, tail wagging and ears pricked forward, in a posture that says: "Let's play!" Human children have their equivalent - a giggle that takes them into the world of make-believe.

Playful dolphins: these very social marine mammals seem to follow ships "for fun"

Feline fun-time

Many of a playing kitten's actions can be interpreted in terms of the hunting techniques used by the adult cat. Kittens play on their own, testing themselves, and also in a group, pretending to be the hunter or the hunted.

SWAT
Mid-air swipes require good eye-paw coordination if they are to connect with a moving target. This type of movement is made as a hunting cat claws a low-flying bird, or clouts a mouse that has leapt into the air.

SCOOP
The kitten tries to put its paw, sole up, under the ball and pick it up or flip it over. The ball does not turn over and so intrigues the animal. Adult cats scoop up small prey, including fish, using this type of movement.

TEASE
Even some adult cats "play" with a small animal before finally killing it. The squeaks of a captive shrew or flutterings of a grounded bird seem to provide entertainment, but the significance of this behaviour is not clear.

POUNCE
The "mouse pounce" is one of the most characteristic cat actions. Other hunters use it too, such as the foxes. The aim is to come down suddenly and silently on the victim's back, away from its teeth and claws, and then, before it has time to resist, sink the teeth into its neck. In this case, the mother's tail acts as the mouse.

Who's the boss?

Dogs - wolves, jackals, dingos and all domestic dogs - are mostly pack animals. Puppies playing together gradually develop many of the social signals they will use when adult, to keep pack hierarchy and organization.

MAN'S BEST FRIEND
Most dogs behave towards their owners much as they would towards a dominant member of their "pack".

TESTING TIME FOR TEETH
The brown puppy, playing on his own, is biting and gnawing the ring as though it was a bone, testing teeth and strengthening jaw muscles.

TUSSLE BETWEEN THE TWO
The black-and-white puppy appears on the scene and wants to join in the game. The two brace themselves and pull hard. This type of action is seen in many pack-hunters, when they cooperatively bring down a large prey such as a deer. It may also occur when two animals are in dispute over a piece of the prey.

TAIL-BITING
The tug-o'-war goes on until the bigger brown puppy gets the ring from his brother but, having got the toy, he soon tires of it and bites his brother's tail instead. They are still playing, but the bite is hard enough to mean business.

I'M THE BOSS!
The puppies now romp together, snarling and biting in a noisy playfight. Suddenly the black-and-white puppy nips too hard. The brown puppy gets cross and the play turns into a power struggle. Muzzles wrinkle, teeth are bared and the puppies glare at one another. The stronger brown pup stays on top to show who is the boss. The black-and-white puppy rolls on his back as a sign of submission to his stronger brother.

Keeping clean

THE MAMMAL'S FURRY COAT is fine for keeping the animal warm and dry. But it is a dirt-trap, and a paradise for parasites, which feed on shed skin or the nutritious blood flowing just beneath. Licking, scratching, combing, shaking, wallowing, bathing, rubbing, picking, and nibbling are some of the techniques that mammals use to maintain personal hygiene, minimize the risks of disease, and ensure any wounds are kept clean as they heal. Many animals clean themselves, but "social grooming," where one individual cleans another of the same species, is widespread among mammals. Social grooming has several functions. One of course is cleanliness: a helper can more easily reach those awkward places such as the neck and back. Another is social organization: dominant individuals may demand to be groomed by lower-ranking ones. Grooming also spreads the group's scent to all members, allowing them to identify each other and pick out intruders.

YOU SCRATCH MY BACK...
Social grooming in baboons not only keeps these mammals clean, it also establishes each animal's position in the group's hierarchy.

ATTENTIVE MOTHER
Young mammals cannot clean themselves, especially baby rodents, which are born naked and helpless (p. 32). This mother hamster is licking her offspring clean. Its fur must be kept clean and dry; if it gets damp, the fur lies flat and cannot trap body heat, which for such a small animal means rapid cooling and the risk of hypothermia.

The neck and top of the head are cleaned by the front paws

NOT A "DIRTY RAT"
Even a "dirty" animal such as a rat spends much time grooming its fur and cleaning its skin. Rats kept domestically are exceptionally clean and make fine pets. The teeth act as a comb to un-snag and brush out the fur, while teeth and claws scratch off lice and dead skin. Wild rats do carry parasites, especially fleas. In 1346-9, rat fleas that bit humans spread the bacteria that caused the bubonic plague - the Black Death - in which perhaps half the people in Europe died.

Paws can reach difficult places

Teeth are used to "comb" fur

The rat finds a safe, quiet place in which to groom - it does not want to be caught by a predator. It reaches round to its back and flanks, cleaning by nibbling and combing

DUST BATH

Some mammals, such as elephants, use the "dry-shampoo" technique, also employed by many birds to keep their feathers in tip-top condition. This is the dust bath, where dust is kicked or thrown onto the body and then rubbed and scratched and shaken, to dislodge dirt and loosen clinging parasites. The chinchilla maintains its luxurious fur in this way, there being plenty of dust in the rocky Andes mountains of South America. Time and energy spent grooming is worth it: the fur protects the chinchilla from the bitter mountain cold and wind.

Long, deep fur

Chinchillas rolling in dust bath

The underparts, which rub the ground and pick up dirt, are cleaned by the mouth and front paws (compare the cat's method of grooming on the next page)

Rat bends double to clean its underside

HELPFUL FRIEND

Creases in the rhino's thick skin are ideal hideaways for ticks and other parasites. The ox-pecker helps its huge host by picking out and eating the parasites. The bird gets a meal, the rhino receives a clean-up - a biological relationship called "symbiosis."

At the cat-wash

Many people are intrigued as to why the typical domestic cat seems to have so much time to clean itself. It is partly because the pet cat has no real need to spend time hunting - food is provided. Cats may groom when they have nothing better to do. They also groom as a "displacement activity." A cat that has pounced and missed a bird may sit and wash itself, allowing the awkward situation to pass.

TWO HORSES ARE BETTER THAN ONE

Mutual grooming in horses helps remove lice and ticks from difficult-to-reach parts such as the withers (shoulder point) and top of the tail. One horse can scratch and rub itself on a favourite post, but a helper improves efficiency. One horse approaches the other, mouth slightly open to signal a grooming session. The pair then stand neck-to-neck or head-to-tail and nibble each other for 5-10 minutes. In a herd, horses develop special grooming friends.

Supple spine allows cat to bend forward

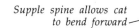

Leg outstretched for balance

U-TURN

The cat's supple, agile body allows its mouth to reach most parts easily. Lying down is more relaxing than balancing on three legs while the fourth is being washed. Paws are especially important, since if they become injured or infected the cat's mobility is reduced, and with it (in the wild) the ability to obtain food. The sole pads are freed of dirt and the claws exposed and examined to check for still-stuck bits of prey.

THAT SHOW-RING SHINE

Horses remove parasites and loose hair themselves, but human preference at the horse show is for a squeaky-clean coat that outshines the opposition. Horses probably judge each other on different points.

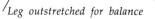

Front paw moistened with saliva used to clean behind ears

THE FACE-RUB

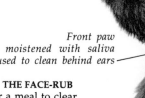

Many mammals (including humans) "lick their lips" after a meal to clear bits of food from around the mouth. The rest of the face cannot be reached so easily by the all-important tongue, carrying saliva that moistens and removes dirt. So the cat licks saliva onto its front paw, wipes and rubs the paw over neck, ear, eye, and whiskers to clean them.

PAW-LICK
People who have been licked by a cat recall the harsh, sandpaper-rough tongue, with its tiny "bristles" (papillae). The human tongue bears these, but they are less stiff and brush-like. The tongue "combs" the fur, while the cat's small incisor teeth work like tweezers to nibble and pick off broken and loose hairs, dead skin, clinging dirt, and parasites. As grooming continues, the cat becomes more contented and relaxed. The front paws are usually "washed" towards the end of the session, since they have been used to clean other areas, such as the face.

Cat lies at full stretch as it becomes contented

Rasping tongue "combs" fur

Fur is still damp from grooming

Back leg sticks up behind head

THE END RESULT
One clean cat. The fur is still slightly damp from saliva, but it soon dries and fluffs up. Grooming helps to spread waxy and oily skin secretions over the skin and hair, forming a semi-waterproof barrier that repels germs. Frequent use of strong shampoos tends to wash these natural secretions from human hair. Mammals should be clean, but not too clean!

REAR-END ACCESS
This is one of the most characteristic poses of a cleaning cat, with one leg held up in the air to gain access to the abdominal and anal areas. Any bits of faeces and dampness from urine are removed.

How to deal with a meal

A LARGE, "COLD-BLOODED" ANIMAL such as a snake may go for weeks without eating. But mammals, being active and "warm-blooded," need lots of energy to keep them going. Energy, as well as the raw materials for growth and reproduction and body maintenance, comes from food. Feeding is therefore vital to life. In modern society, humans spend relatively little time out hunting for food. It may take what seems like all day to go round the supermarket, but we have lost sight of how most wild mammals build their daily routine around finding enough to eat. One of the reasons for the mammal's high energy requirements is the ability to be active in cold conditions, when the "cold-bloodeds" are chilled and slow. This may be why much mammalian food-hunting is done at dawn and dusk, before the heat of the day allows reptiles, insects, and other cold-blooded prey to warm up and dart away. The smaller the mammal, the more feeding it has to do, since small bodies have proportionally more surface area than large ones, and so lose heat at a greater rate. In cooler climates, the smallest mammals have only just enough hours in the day to feed themselves. Shrews do little else except feed in a frenzy, then rest and digest, then feed again. They eat their own body weight in food each day, and can starve to death in only three hours. At the other end of the meat-eater scale, the lion needs only the equivalent of about 1/40th of its body weight in food each day. Mouths and teeth give evidence as to the types of food eaten (p. 50); claws are also good clues (p. 58).

Three-course meal: house mice make short work of jugged cream, bulbs, and candle wax during their nightly trips to the kitchen

TREE-TOP TONGUE
The giraffe's long, dark tongue stretches upwards to add another 30 cm (1 ft) or so of height to this the tallest of land mammals. Vegetation more than 5.5 m (20 ft) high can be cropped by a large male giraffe. The tongue grasps leaves and twigs and pulls them within reach. The canine teeth have two deep grooves to strip leaves from their twigs.

Chipmunk feeding on nuts

HAND-TO-MOUTH
The chipmunk holding food in its hand-like forepaws is a common sight in eastern North America. These naturally curious members of the squirrel family frequent picnic sites and parks, in the hope of leftover titbits. The chipmunk manipulates food in a most efficient manner. As it feeds, it rotates food items quickly, scrabbling off loose bits and testing with the teeth to find the weak point where nuts can be cracked. Like many other rodents, it carries surplus food in cheek pouches back to its burrow (p. 52).

Front paws are used to turn food

GRAIN DRAINS

Apart from ourselves, house mice are probably the world's greatest graminivores (grain-eaters). Even in the wild these small rodents have an amazingly varied diet and tackle seeds, fruits, leaves, shoots, and other plant matter, and also insects and other small creatures. In and around human habitation they become still more unfussy. They have eaten bread, paper, string, butter, soap, candle wax (see opposite), and other waxy or fatty substances, plus of course the famous cheese used to bait mousetraps. House mice have even been known to invade butchers' cold stores and feed on chilled or frozen meat. The mouse gnaws and chisels at the food with its long, sharp front incisors typical of the rodent group (p. 50); it usually holds small items in its front paws. Its lower incisors leave two characteristic grooves.

Seeds are held in fore paws

House mice feeding on grain

Heading off to a secret store . . . (p.52)

Mice sit on back legs when using front paws to hold food

Mouse is alert to danger even when feeding

CLAWS FOR EVERY OCCASION

The Malayan sun bear, like most of its relatives (p. 50), is an all-round feeder (omnivore). It is the smallest bear, and its light weight and long, curved claws (relative to other species) enable it to climb well and also hook ripe fruit from branches. It also tears bark from trees with its claws to uncover grubs and the nests of termites and bees.

THE RIGHT TOOL FOR THE JOB

The chimp's brain takes over when its physique fails. Faced with termites deep in a strong nest, our closest relative takes a stick and dips it into the hole. The termites grab it, and the chimp fishes them out and licks off a tasty meal. A few mammals, and some birds, are known to use tools in this fashion.

FISH SUPPER

The otter rarely eats its catch in the water. It comes to the bank and holds down the slippery meal with its front feet while tearing at the flesh with its sharp, spear-like canine teeth (p. 50). Otters also eat small mammals, birds, and frogs.

Grippers and grinders

MAMMALS, being warm-blooded and generally active, need to take in plenty of food to provide sufficient energy for life. The jaws and teeth are at the forefront of feeding: they grasp the food, cut it into smaller pieces if necessary and do some preliminary crushing and grinding before it is swallowed. The basic structure of the mammalian tooth is a soft inside containing nerve and blood vessels, overlaid by tough dentine, and covered by enamel. Yet from this simple plan, mammals have evolved a huge variety of choppers, snippers, impalers, shearers, grippers, grinders, and many other tooth shapes. Teeth are especially important in studies of mammal evolution, because they are often

Canine tooth

Molar tooth

well preserved as fossils and so are one of the few ways of comparing extinct species with living ones (pp. 12-15).

The male narwhal's 3 m (10 ft) "tusk" is an overgrown left tooth. Its function is not certain - it may be a male symbol to win over females

THE ADAPTABLE BEAR
Bears are classed in the Order Carnivora (p. 8), but in reality some species eat a variety of foods in season, from fish, rodents, and young deer to buds, fruits, and berries - and the legendary honey.
The bear's teeth are adapted accordingly - pointed incisors and canines for the meat and grinding molars for plant material.

Honeycomb

Fish flesh

MOUTHFUL OF GRASS
The horse's teeth are in two main groups. Small, sharp ones at the front (incisors work with the lips and act as croppers to snip off grass. Large, flat teeth at the back (molars) grind the goodness out of the grass.

PANDA PUZZLE
The giant panda has long perplexed experts. Its general body structure indicates that it belongs to the Order Carnivora (meat-eaters), yet its diet is principally herbivorous - it eats mostly bamboo, although it will also eat insects, small mammals, and carrion. Recent evidence suggests that its closest relatives are probably the bears.

Very long incisor teeth are orange in colour

Small canine tooth ("tush"), only in male horse

Incisors

Panda lower jaw

Large, flattened tooth for chewing plants, typical of a herbivore

Canadian beaver lower jaw

EVER-GROWING INCISORS
The beaver is a member of the Order Rodentia (p. 9). Rodents have long, chisel-like incisors (front teeth) specially designed for gnawing. These teeth are continually worn down as they chip and chisel at wood and other tough plant food, so it is as well that they grow all the time - otherwise the beaver would starve.

Large molar tooth

Bamboo - the panda's main food

Grass

Large canine tooth typical of carnivore

Bark and buds - the beaver's food

Horse lower jaw

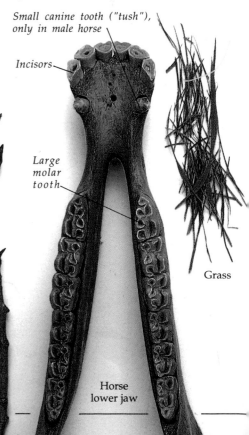

CARNASSIAL POWER

Jackals are often considered scavengers, cleaning up the leftovers at a lion's kill. But they hunt too. The ridged carnassial teeth near the jaw joint can shear skin, gristle, and bone.

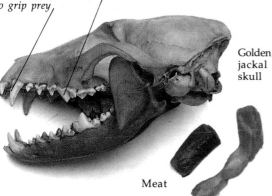

Canine tooth to grip prey

Carnassial tooth to cut meat

Golden jackal skull

Meat

NO TEETH

The long-beaked echidna eats small worms and insects. It has no teeth - the prey is taken in by a sticky, spiny tongue and mashed between the rough back of the tongue and the roof of the mouth.

Tongue housed in long tube

Long-beaked echidna skull

THE UNUSUAL AARDVARK

Africa's aardvark is unusual in many ways. It only has back teeth, and these have no enamel. They do little chewing since ants and termites collected by the sticky tongue are crushed in the specialized stomach.

RAZOR SHARP

The hedgehog has small, sharp teeth to chew up its diet of caterpillars, grubs, and beetles.

Hedgehog skull

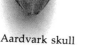

Peg-like teeth

Teeth are very sharp

Aardvark skull from below

Fringed plate of baleen

SLIPPERY CATCH

The dolphin is a piscivore (fish-eater). Its beak-like mouth bears teeth that are small, sharp, and all the same - an excellent design for holding slippery fish or squid.

Dolphin lower jaw

Lower jaw is long and delicate

Teeth are all the same

Baleen is made of fibres

KRILL COMB

Baby baleen whales growing in the womb have tiny teeth for a while. But the plates of baleen, or "whalebone" take over, developing from the roof of the mouth. The whale gulps sea water into its mouth, forces it out through the baleen sieve, and then "licks off" the krill (see below) and other small creatures and swallows them.

Diagram to show rows of baleen in whale's mouth

Dolphin's dinner - squid and fish

A tamandua (collared anteater) forages for ants and termites with its long, sticky tongue; it has no teeth (left)

Crab-eating seal skull

Notched teeth for sieving krill

THE CURIOUS CASE OF THE CRAB-EATER

The crab-eating seal of the Antarctic does not actually eat crabs, it eats krill. Look closely at the intricate teeth of this seal and you will see how it can sieve tiny shrimp-like krill from the sea-water under the pack ice.

Krill

Food for later

FEW HABITATS IN THE WORLD provide a constant, year-round supply of food. Our distant ancestors recognized this, and they understood the need to build up a store for later. Planning ahead by planting crops and storing fruits led to the beginnings of agriculture, some 10,000 years ago. Other mammals, however, have probably been saving food in times of plenty, for eating when times are hard, for millions of years. Seeds are a favourite. In a seed, the parent plant has provided a rich store of nutrients that the embryo (baby) plant will rely on when it germinates. The seed is therefore a ready-packed, nutritious meal. In return, the seed-storers help the plant. An animal that buries seeds and then forgets about them has helped the plant to spread. Meat is more of a problem since it tends to decay, but burying is still worthwhile for mammals such as foxes. With its legendary "cunning" the fox does not store all its surplus food in one place. It makes several stores in different places, so that if another animal discovers one store, it does not lose the lot.

A cheeky way to collect food

The golden hamster is a rodent (p. 9). Like many of its relatives it collects food when this is abundant and "caches" it (stores it away in a hidden place). The hamster's cheek skin is loose and floppy, and forms an expandable pouch in which food is carried. A number of mammals carry food in this way, including the platypus.

2 PACKING THE POUCHES
The nuts are quickly put into the mouth and are then pushed into the the pouches using the tongue. The hamster pauses occasionally to check for danger, then hurriedly continues.

Pouches are beginning to extend

Cheek pouches are empty

Pile of nuts

1 A LUCKY FIND
Although golden hamsters are kept as pets, their pouch-filling behaviour is also shown by their wild cousins, such as the common hamster of Eastern Europe and Central Asia. Here the lucky hamster has found a pile of nuts.

PACA'S PACKAGES

The paca is a nocturnal rodent about the size of a smallish dog, that lives in northern South America. Its square-headed look is due to its curved, bowl-like cheek bones, once thought to be used for storing food. In fact, their exact function is not clear - one theory is that they are used to amplify the sound that the paca makes.

Paca skull

Nasal passage

Front teeth

Eye socket

Expanded cheek bones

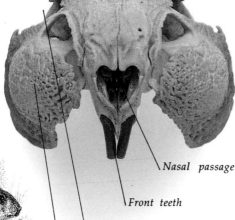

Food for the future

Mammals use many different methods, and go to great lengths, to store the energy and nutrients that a meal represents. The methods have evolved in response to availability of food in the habitat.

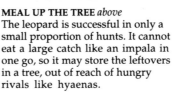

MEAL IN THE TREE
The red fox buries extra food and then returns to it later. But not always successfully - another creature may find it, or the fox may forget where it put it.

MEAL UP THE TREE *above*
The leopard is successful in only a small proportion of hunts. It cannot eat a large catch like an impala in one go, so it may store the leftovers in a tree, out of reach of hungry rivals like hyaenas.

WINTER WARMTH AND ENERGY *right*
The dormouse feeds greedily on autumn fruits and builds up stores of fat under the skin. This provides enough energy for a half-year of hibernation.

Pouches are now full

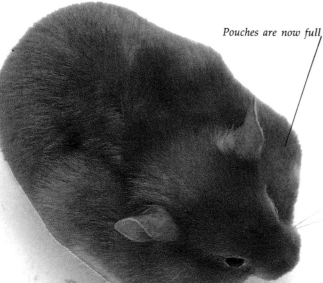

3 SHOPPING BAGS FULL
Like a human shopper staggering home from the market with a heavy bag in each hand, the hamster has packed its cheek pouches to bulging with nuts. Now it is time to leave the feeding area, which is exposed and therefore hazardous for such a small, relatively defenceless rodent.

4 FROM BAG TO BURROW
When the hamster reaches the safety of its burrow, it "unpacks" its pouches. The front paws are used like hands to push and "massage" the food out of the pouch, into the creature's underground larder. In the wild, a single hamster has been found to collect more than 60 kg (132 lb) of nuts and other food (equivalent in weight to an adult human).

Hamster uses front paws to get nuts out of pouches

At home in a nest

Harvest mouse nest built on cereal stalks.

NESTS OF VARIOUS SORTS occur throughout the animal world. We are familiar with birds' nests, and some of the busiest and most elaborate builders are insects such as termites. But mammals have a fair share of species that make conspicuous nests in the open, as well as the many species that nest in burrows (p. 56). They include squirrels in Europe, pack-rats in North America, Karoo rats in Africa, and bandicoots in Australia. One of the most extraordinary mammalian nest-makers is the stick-nest rat, a rabbit-sized native of Australia. This very rare rodent makes a strong, interwoven pile of branches, twigs, and even stones, 1 m (3 ft) high and 2 m (6 ft) across. It lives in the rocky southern lands, where digging is difficult, and the nest probably gives protection against predators. Sadly, this rat has died out on the mainland, and only an island colony remains off the south coast. It seems it was never common, and both Aborigines and Europeans hunted it.

THE GREY IN ITS DREY
A winter walk through woodland in Europe, with trees bared of foliage, may reveal soccer-ball-sized bundles of sticks wedged into the forks of trees. These are dreys, the homes of grey squirrels. Some will be old and deserted, and others will be flimsier summer dreys, not used in winter. But a few dreys will each hold an occupant like this one which, in winter, is not hibernating but probably sleeping. Squirrels are active (mainly at midday) throughout the winter and can only survive a few days without food. They stay in the winter drey at night and in very bad weather. The drey is a tangle of twigs and sticks, some with leaves still attached, and is lined with bark, grass, and other bits and pieces gathered by the owner. This drey is about 45 cm (18 in) in diameter, with an internal chamber 30cm (1 ft) across. Baby squirrels are born in special nursery dreys in spring.

Things that a drey might be made of

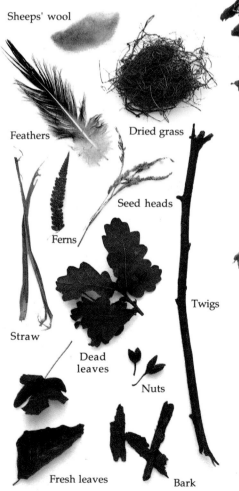

Sheeps' wool

Feathers

Dried grass

Ferns

Seed heads

Straw

Twigs

Dead leaves

Nuts

Fresh leaves

Bark

With straw held in its mouth, gerbil runs back to nest site

SHREDDING THE BEDDING
In the wild the gerbil, a small, desert-dwelling rodent, digs burrows away from the heat and dryness, and lines them with shredded plant matter. In captivity, the results of digging behaviour may be prevented by the cage. The lining is more successful: here the animal collects suitable material.

DAY ONE
The "raw straw" that was supplied to two gerbils before they woke up in the evening.

DAY TWO
A night of shredding with the teeth produces a partly made nest.

DAY THREE
More shredding, and the nest "fluffs out" and takes shape.

WHAT'S IN A DREY?
Grey squirrels tend to take any suitable ingredients for their dreys. In towns, where human litter is more prevalent, they have been known to incorporate plastic bags, drinking straws, and newspaper into the drey.

Grey squirrel drey cut in half to show inside

Winter drey is well-built, unlike flimsy summer drey

Inner, cosy lining

Outer layer of twigs and leaves

Squirrel turns round and round to shape drey

Drey is made in fork of a tree

Life underground

PRAIRIE, PAMPAS, SAVANNA, STEPPE, and other types of
grassland are some of the best areas to find burrowing
mammals. As there are few trees, and so little shelter, the
main refuge is underground. North American prairie dogs
and ground squirrels, South American viscachas and maras,
African root rats and mole-rats, and Asian sousliks and
gerbils all tunnel in grasslands. They gain safety, a place to
build a nest to rest and breed, and shelter
from the hot sun and cold wind. But most of
them have to emerge at some time, since they
are chiefly herbivores, and plants do not grow
underground in the dark. More specialized
feeders, such as blind mole-rats, gnaw roots,
bulbs, tubers, and other underground plant
parts, and can stay permanently below the
surface. Then there are the insect-eaters,
such as the moles . . .

UNDER THE MOLE'S MOUND
Champion among mammal burrowers is the
European mole, which lives, breeds, sleeps, and eats
underground. A few mounds of fresh soil in a meadow
are the only sign of a complex system of burrows and
chambers 1 m (3 ft) or more beneath, and possibly
stretching 100 m (300 ft) in length. The size of the
burrow is governed largely by richness of soil. In old
pasture, with lots of earthworms and insects, a mole
would have to burrow less than in poorer, stonier, or
sandier soil. Most of the food comes from the "mole
patrol" as the animal wanders its tunnels, main-
taining them and eating creatures that have fallen
in from the walls.

BANKSIDE FRONT DOOR
The platypus retires
to a burrow in the
riverbank after feeding.
Resting burrows are
usually under tree roots,
and are a few metres
long. The breeding
burrow is much longer,
and as the pregnant
female enters she
blocks it with mud at
intervals, to protect
against floods and
intruders and to keep
herself warm. At its
end, in a grass-lined
nest, her eggs
are laid (p. 31).

THE SNOW DEN
As the Arctic nights
become almost contin-
uous in winter, the
mother polar bear digs
a den in drifted snow.
About one month later
her cubs are born, and
she stays with them
and suckles them
for around three more
months. Spring arrives
and the family emerges -
cubs well-fed and
chubby, but mother
thin and hungry, eager
for her first seal meal
for four months.

KEY TO THE MOLE BURROW

1 The fortress - not an ordinary molehill, but the larger, more permanent mound above the mole's main nest

2 The nest - the female breeds in spring, giving birth to about four pink babies in a nest lined with grass, leaves, and other soft material

3 The mother mole - gathering nest lining is one reason for the mole's hazardous journeys above ground, usually at night

4 Surface run - some tunnels run just below the soil surface

5 Friend or foe? - European moles are solitary creatures. An individual that blunders into another's tunnels is usually chased away except in early spring when it could be a potential mate

6 The pantry - moles bite off worms' heads and store them in an underground larder, especially in autumn

7 Criss-cross tunnels - these run at all angles

Key diagram

How many toes?

THE ORIGINAL MAMMALS probably walked on all four legs, on paws with five "fingers." Today, there is almost every variation imaginable. The horse, a sizeable mammal, walks on its "tip-toes," and has only one toe on each foot. Small mammals such as the shrew still have all five digits. In general, a mammal with long limbs is a swift mover, while short limbs indicate strength and perhaps digging ability. Gazelles and antelopes have ultra-slim limbs for speed, while seals and bats both possess large, finger-supported limb surfaces to push aside water and air respectively. Claws, nails, hooves, fleshy pads, and other structures tip the toes.

Shetland pony forelimb skeleton

— *Lower arm bone*

— *Wrist bones*

TWO-TOED WALKING
The gazelles are even-toed hoofed mammals (p. 10), and these dainty feet allow them to run at great speed.

Soemmerring's gazelle forelimb skeleton

— *Lower arm bone*

DISAPPEARING TOE
The tapir is an odd-toed hoofed mammal, like the horse (p.10). Odder still is the fact that its front feet have four toes while its back ones have three. The fourth toe is smaller than the others and does not touch the ground, except when it is very soft.

— *Palm bone*

Brazilian tapir forelimb skeleton

— *Lower arm bones*

— *Wrist bones*

Wrist bones

Finger bones

— *Palm bones*

Wrist bones

Canon (palm) bone

Finger bones

PENTADACTYL PLAN
The basic mammalian limb ends in five digits, like our own hands and feet. Many rodents, primates, and carnivores have kept this "penta-dactyl" design. The hoofed mammals have lost different digits in different groups. Each bone, or set of bones, in the limb is represented by the same colour throughout. (Names in brackets refer to the equivalent bones in the foot and lower leg.)

Key to coloured bones (based on human hand)

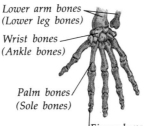

Lower arm bones (Lower leg bones)

Wrist bones (Ankle bones)

Palm bones (Sole bones)

Finger bones (Toe bones)

INSIDE THE HOOF
The zebra's hoof is made of hard, protective horn with a shock-absorbing pad of fat (the plantar cushion) between it and the toe bones.

Section through zebra's hoof

Finger bone

Position of plantar cushion

Toe bone

Outer hoof made of horn

ONE-TOED WALKING
The slim limb bone of the horse has been modified by evolution to leave only one toe, the 3rd (middle) or pastern (p. 13). This is joined to a long, thick canon bone which represents the fusing of the palm bones. The whole design does away with the numerous fingers and toes, with their heavy muscles and joints. It combines lightness with strength, especially toward the end of the limb, giving the horse its speed.

SAND-SHOED MAMMAL
The camel, another even-toed hoofed mammal, has enlarged, flexible foot pads that spread its weight well over the soft desert sand.

FLEXIBLE FEET
The dinky feet of the rock hyrax bear flat-tened nails, not true hooves: four at the front, three at the back.

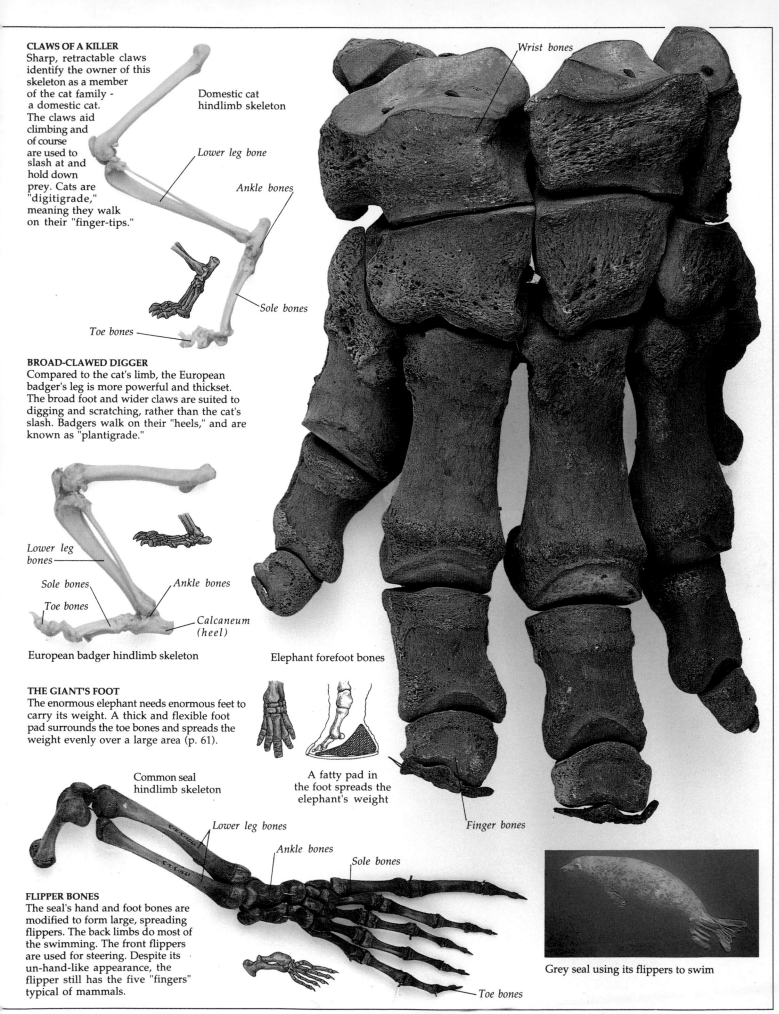

CLAWS OF A KILLER
Sharp, retractable claws identify the owner of this skeleton as a member of the cat family - a domestic cat. The claws aid climbing and of course are used to slash at and hold down prey. Cats are "digitigrade," meaning they walk on their "finger-tips."

Domestic cat hindlimb skeleton

Lower leg bone

Ankle bones

Sole bones

Toe bones

BROAD-CLAWED DIGGER
Compared to the cat's limb, the European badger's leg is more powerful and thickset. The broad foot and wider claws are suited to digging and scratching, rather than the cat's slash. Badgers walk on their "heels," and are known as "plantigrade."

Lower leg bones

Sole bones

Toe bones

Ankle bones

Calcaneum (heel)

European badger hindlimb skeleton

THE GIANT'S FOOT
The enormous elephant needs enormous feet to carry its weight. A thick and flexible foot pad surrounds the toe bones and spreads the weight evenly over a large area (p. 61).

Elephant forefoot bones

A fatty pad in the foot spreads the elephant's weight

Common seal hindlimb skeleton

Lower leg bones

Ankle bones

Sole bones

FLIPPER BONES
The seal's hand and foot bones are modified to form large, spreading flippers. The back limbs do most of the swimming. The front flippers are used for steering. Despite its un-hand-like appearance, the flipper still has the five "fingers" typical of mammals.

Toe bones

Wrist bones

Finger bones

Grey seal using its flippers to swim

Domestic cat

Tracks and trails

WALKING THROUGH any wild place, we are aware of many animals. Birds fly above, insects buzz from one flower to another, and fish rise to snatch food from the water's surface. But where are all the mammals? With their swift and active habits and their keen senses (p. 16) they make themselves scarce, fearing the large creature blundering past. Others, being nocturnal, are well hidden and asleep. Although we are mammals ourselves, we seldom have the time and patience to glimpse our wild relatives. Often, we only know of their presence from the tracks and signs they leave behind; footprints and belly- or tail-drags in the ground, leftover bits of food with teeth marks, droppings, burrow entrances with dug-out soil, bits of hair caught on twigs and snagged on thorns, and cast-offs such as antlers (p. 62). The footprints shown here are actual size, and actual prints made by the walkers themselves: real and "messy," not cleaned and tidied up. The prints were made by encouraging the animals (by bribing with food) to walk on a pad of non-toxic ink and then across the paper. Claw marks do not show up using this technique, but they will be found in trails in soft mud or snow. In a trail, the spacing of the prints and the depths of their impressions allows us to work out whether the animal was running or walking.

Fore-print

Toe pad

Intermediate pad

Hind-print

Fur on sole of foot

DIGITIGRADE KITTY
The domestic cat is digitigrade (a toe-walker, p. 59) and its toe pads are well separated from the main three-lobed sole, or intermediate pad. There are no claw marks: the claws are kept sharp in their sheaths until needed. Neither is there a mark from the innermost (1st) toe on each front foot, which is too high to leave a mark. Hence both fore- and hind-prints are four-toed and roughly the same.

Fore-print

RUN, RABBIT, RUN
When sitting or hopping slowly, the rabbit's hind foot leaves its characteristic long imprint compared to the more circular front foot. But when running the difference is less obvious, since the animal tends to place only the tips of its hind feet on the ground.

Hind-print

CLOVEN HOOVES
Animals that like mud provide plenty of prints in the soft ground. And the heavier the animal, the better. A half-tonne wallowing buffalo left this clear "cloven-hoofed" print, indicating it is an artiodactyl, or even-toed hoofed mammal (p. 10).

Feet are covered with fur - no pads show

Rabbit

Print of fur would not show up in snow

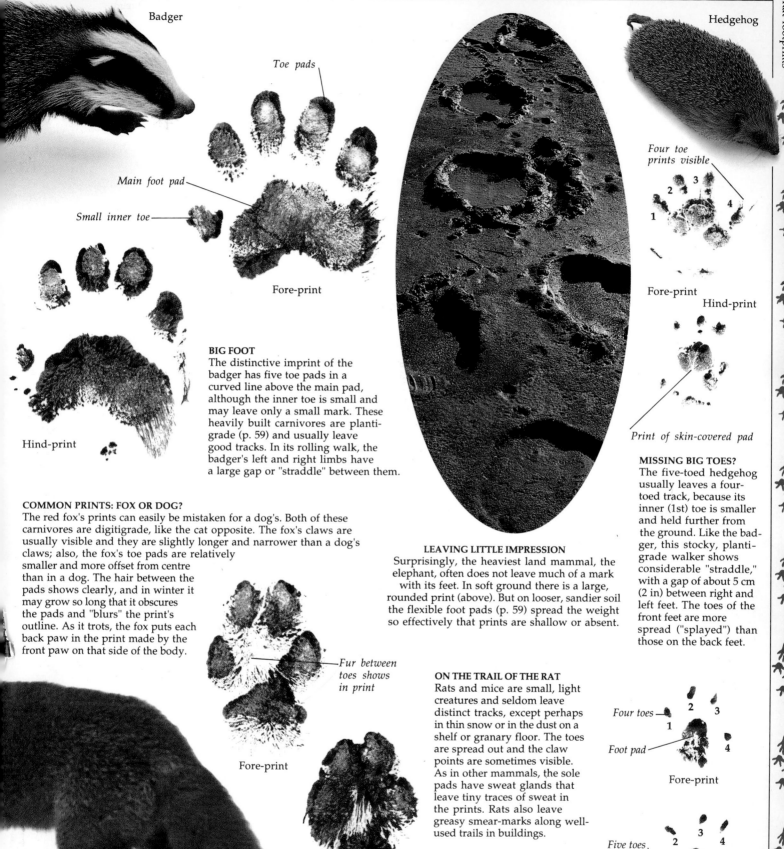

Badger

Hedgehog

Toe pads

Main foot pad

Small inner toe

Fore-print

Hind-print

Four toe prints visible

Fore-print

Hind-print

Print of skin-covered pad

BIG FOOT
The distinctive imprint of the badger has five toe pads in a curved line above the main pad, although the inner toe is small and may leave only a small mark. These heavily built carnivores are plantigrade (p. 59) and usually leave good tracks. In its rolling walk, the badger's left and right limbs have a large gap or "straddle" between them.

COMMON PRINTS: FOX OR DOG?
The red fox's prints can easily be mistaken for a dog's. Both of these carnivores are digitigrade, like the cat opposite. The fox's claws are usually visible and they are slightly longer and narrower than a dog's claws; also, the fox's toe pads are relatively smaller and more offset from centre than in a dog. The hair between the pads shows clearly, and in winter it may grow so long that it obscures the pads and "blurs" the print's outline. As it trots, the fox puts each back paw in the print made by the front paw on that side of the body.

Fur between toes shows in print

Fore-print

Hind-print

Red fox

LEAVING LITTLE IMPRESSION
Surprisingly, the heaviest land mammal, the elephant, often does not leave much of a mark with its feet. In soft ground there is a large, rounded print (above). But on looser, sandier soil the flexible foot pads (p. 59) spread the weight so effectively that prints are shallow or absent.

MISSING BIG TOES?
The five-toed hedgehog usually leaves a four-toed track, because its inner (1st) toe is smaller and held further from the ground. Like the badger, this stocky, plantigrade walker shows considerable "straddle," with a gap of about 5 cm (2 in) between right and left feet. The toes of the front feet are more spread ("splayed") than those on the back feet.

ON THE TRAIL OF THE RAT
Rats and mice are small, light creatures and seldom leave distinct tracks, except perhaps in thin snow or in the dust on a shelf or granary floor. The toes are spread out and the claw points are sometimes visible. As in other mammals, the sole pads have sweat glands that leave tiny traces of sweat in the prints. Rats also leave greasy smear-marks along well-used trails in buildings.

Four toes

Foot pad

Fore-print

Five toes

Hind-print

Brown rat

Mammal detective

Indian trackers rely on their detective work for food

FOR MOST PEOPLE TODAY, contact with the natural world is limited to the garden or park, or an occasional walk in the woods. This unfamiliarity with nature breeds a type of "blindness:" when out on a walk we look, but we don't know exactly what for. Yet there are still groups of peoples around the world who live with nature, in the manner of our ancestors. We can only wonder at their knowledge and experience when it comes to "detective work." The merest hint of gnawing or a dropping is quickly identified since it is important - it could lead to meat for food, bones for tools, and skins for clothes and shelter. Yet anyone can learn. It's a question of having the time, and needing the knowledge.

Long-lasting bones

Bones, teeth, horns, antlers, and other hard parts of the mammal body tend to persist long after the flesh and soft organs have been eaten or rotted away. To the trained eye, a crack or dent in a certain place can indicate the manner of death. Wear on the teeth may show that the owner was old and weak, and perhaps died from disease.

HORNED SKULL
The "brain-box" (cranium) of the skull is designed to protect the brain within, and even on this old sheep it has not been broken. Small carrion-feeders crawled inside and picked the skull clean.

Round hole gnawed out by dormouse

Fascinating dung

Many mammals have regular defecation stations, and the droppings are often used as territorial markers, as when an otter leaves its "spraints."

Rabbit droppings

SQUIRREL SIGNS
Squirrels strip the scales from pine cones to reach the nutritious seeds sandwiched inside.

Squirrel-gnawed pine cones

NUT-CRACKERS
The hard shell of the hazel cob (nut) is a challenge, but the delicious kernel inside is worth it. Different mammals tackle the shell in characteristic ways.

Nut split cleanly in two by adult squirrel

RABBIT "PEAS"
Rabbits use their droppings to scent-mark territory.

Teeth at work

Rodents are the champion gnawers. Even when not feeding, they gnaw experimentally at many different materials using their chisel-shaped incisor teeth (p. 50).

Irregular hole in side, the work of a yellow-necked mouse

Shells gnawed by rat

SNAILS UNSHELLED
A brown rat on the beach neatly gnawed these winkle shells to eat the occupants.

Roe deer droppings

Electrical cable gnawed by rodent

DEER DROPPINGS
Deer eat lots of low-nutrient food and so leave large amounts of droppings.

POWER CUTS
Rats and mice may gnaw at electrical cables to find out what is inside. This can have consequences. Sometimes the animal is electrocuted. Fires and power cuts have been started by such "innocent" rodent behaviour.

Lower jaw of gnawing
animal - long incisors

Herbivore's tooth -
flat, grinding top

Lower jaw of carnivore - carnassial tooth

BLEACHED, BEACHED BACKBONE
Cleaned white by the sea, this
fur seal's vertebra (backbone) was
washed up on the appropriately
named Skeleton Coast, Namibia,
South-West Africa. The salty
water has caused chemical
corrosion, dissolving out the
weaker substances to show the
internal structure of bone.

Fur seal vertebra

*Internal channels
in bone can
be seen*

UNREWARDING MOUTHFULS
Jaws and teeth are rarely eaten by a
predator, since the teeth are too
hard and their roots project
into the jawbone.

A NATURAL DEATH?
In urban areas
approximately 50
per cent of fox deaths
each year are caused
by cars. These bones
were found near a main
road. Perhaps the fox
was hit by a car and
crawled away before
dying.

CAST-AWAY
Deer lose or "cast" their
antlers each year, and
grow a new set. The roe buck
uses its antlers in duels
with other males (p. 26)
and also rubs them on trees
during the summer, to mark
his territory.

Roe deer
antler

Pelvis
(hip bone)

*Broken
shaft*

*Point where antler is
joined to the skull*

A PILE OF WINGS
This indicates that a bat is
nearby. They are partial to the
juicy bodies of moths, but allow
the dry wings to drop into a neat
heap below their roost.

Limb bones

Furs on the fence

Barbed wire is the artificial equivalent of the thorn hedge, and just as productive for snatches of fur
snagged from passing animals. The height at which the fur was caught, and the size of the hole through
which the animal pushed, are important clues, as well as the colour and nature of the hairs.

Fox fur

Sheep fur (wool)

Rabbit fur

Index

Acknowledgements

Dorling Kindersley would like to thank:
Jane Burton and Kim Taylor for all their ideas, hard work and enthusiasm.
Dave King and Jonathan Buckley for additional live animal photography.
Daphne Hills, Alan Gentry and Kim Bryan at the Natural History Museum for the loan of the specimens and for checking the text, and Colin Keates for photographing the collections.
Hudson's Bay, London, for loan of the furs.
Will Long and Richard Davies of Oxford Scientific Films for photographing the section through the molehill.
Jo Spector and Jack.
Intellectual Animals, Molly Badham and Nick and Diane

Mawby for loaning animals.
Elizabeth Eyres, Victoria Sorzano, Anna Walsh, Angela Murphy, Meryl Silbert and Bruce Coleman Ltd.
Radius Graphics for artwork.

Picture credits
t= top b=bottom m=middle l=left r=right

Archive fur Kunst und Geshichte, Berlin: 12b; 15b
Pete Atkinson/Seaphot: 59b
Jen and Des Bartlett/Bruce Coleman Ltd: 29b
G I Bernard: 37bl; 60m
Liz and Tony Bomford/Survival Anglia: 33t
Danny Bryantowich: 23bl
Jane Burton: 27m
Jane Burton/Bruce Coleman Ltd: 17b; 18b; 21mr; 27t; 34b; 36b; 46t; 53m;
John Cancalosi/Bruce Coleman Ltd: 27m
Peter Davey/Bruce Coleman Ltd: 49bm

Jeff Foott/Bruce Coleman Ltd: 39t
Frank Greenaway: 63m
David T. Grewcock/Frank Lane Picture Agency: 53mt
Zig Leszczynski/Oxford Scientific Films: 48m
Will Long and Richard Davies/Oxford Scientific Films Ltd: 57
Mansell Collection:19t
Mary Evans Picture Library: 8tl; 16bm; 20m; 26m; 28m; 29m; 31m; 37bl; 46b; 49br; 58m
Richard Matthews/Seaphot: 27t
Military Archive & Research Services, Lincs.: 22b
Stan Osolinski/Oxford Scientific Films Ltd: 44t
Richard Packwood/Oxford Scientific Films Ltd: 16br
J E Palins/Oxford Scientific Films Ltd: 16bl
Dieter and Mary Plage/Bruce Coleman Ltd: 61r
Masood Qureshi/Bruce Coleman Ltd: 45b
Hans Reinhard: 26m

Jonathan Scott/Planet Earth: 53mt
Kim Taylor/Bruce Coleman Ltd: 32b

Illustrations by John Woodcock:
8; 9; 10; 11; 13; 14; 19; 20; 27; 58; 59